# CHAPTER ONE

## INTRODUCTION
## &
## LITERATURE SURVEY

## 1.1   The Central Nervous System (CNS)

The central nervous system consists of two parts, the brain and spinal cord (1). The brain includes :

a.   The two cerebral hemispheres.
b.   The brain stem.
c.   The cerebellum.

The brain stem is a narrow structure in which all the pathways entering and leaving the two hemispheres pass through.  It contains, the centers that control many critical functions like the control of breathing, heart rate and eye movement. The brain stem includes the mid brain, pons and medulla oblongata (2) (figure 1.1 A,B).  Movement and balance are under the control of the cerebellum, a small rounded structure located beneath the cerebral hemispheres.  The spinal cord is of great importance in nervous conduction and reflex activity.  It has a long cylindrical shape and consists of the gray matter which is predominantly composed of nerve cell bodies, the white matter which is mainly composed of myelinated fibers(3). The spinal cord represents the point of exit for nerves on their way out to the muscles they control and the point of entry for sensory fibers returning from the body's sensory organs (1).

### 1.1.1   The Meninges

The meninges are three covering membranes,  that surround the soft brain and spinal cord tissue within their rigid bony encasement (4).  The strong outermost covering is the dura. The innermost is a delicate cellular-fibrillar matter named the pia. Between the dura and pia is the arachnoid, also cellular and fibrillar (figure 1.2). The dura is often termed pachymeninx, while the pia and arachnoid together are called the leptomeninges.  These terms are given because of the relative thickness of the three membranes (4).  The tough and nonelastic properties of the dura, keeps the three components blood, brain and CSF within a space of a fixed volume.  The space between the pia and the arachnoid is called the subarachnoid space (1).

The three meninges surrounding the brain and spinal cord, are separated from one another by spaces (intermeningeal spaces)  containing fluid. The subdural space which is the space between dura and arachnoid, contains a small amount of fluid which is probably similar to the CSF.   The CSF circulates within the subarachnoid space.  It is also present in the cisterns, which are large pocket like areas in the intracranial subarachnoid space (4).

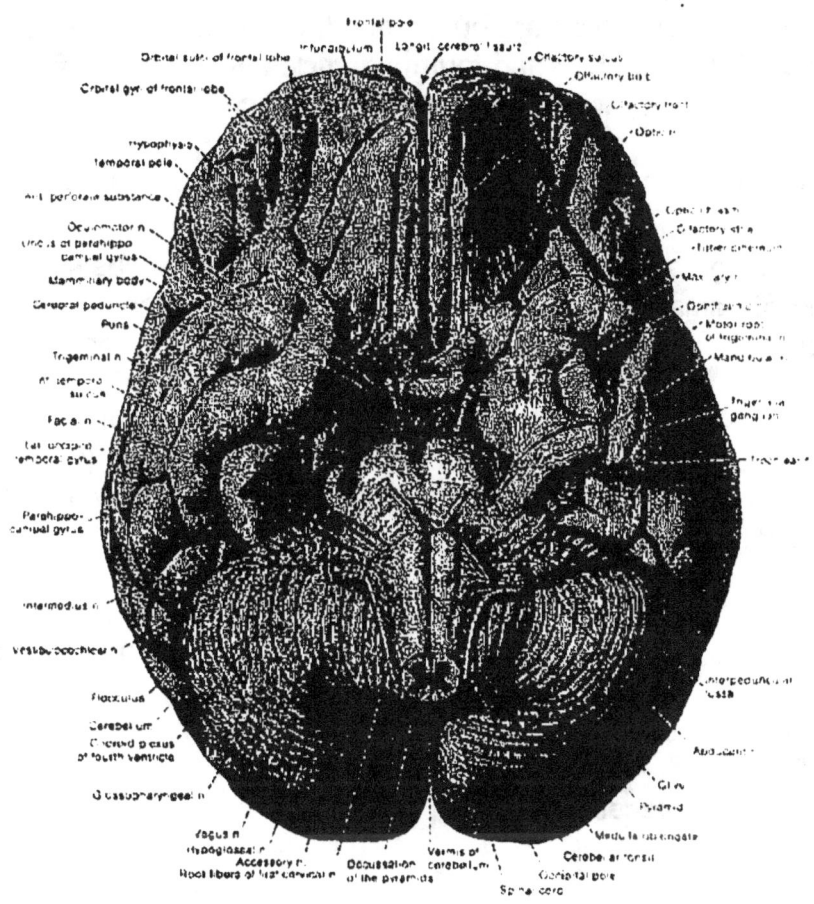

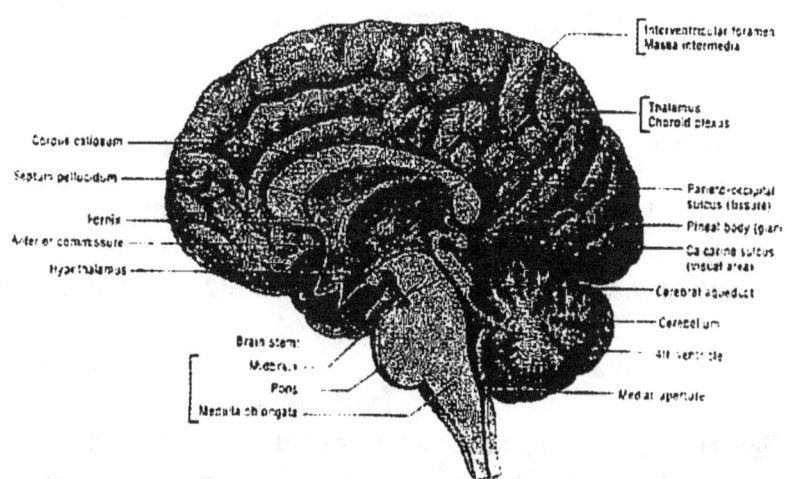

**Figure (1.1). The brain. (A)** : Basal view of the brain revealing the two cerebral hemispheres, the brain stem and medulla oblongata. **(B)** : Median section of the brain. *(Source:Sobotta atlas of anatomy, vol. 1(1983), p.51 & Grants atlas of anatomy, 9th edition (1991), p.473).*

The Ventricular System : It is a connecting system of hollow spaces in which the two cerebral hemispheres are built around. It consists of four ventricles: the two lateral ventricles, the third and fourth ventricles (5). They are filled with CSF and communicate directly with the subarachnoid space (1).

## 1.1.2 Pathological conditions of the central nervous system

Damage of the brain and spinal cord, lead to various symptoms related to the type of damage. Brain and spinal cord damage, could be focal or diffuse. Damage of focal areas, lead to paralysis of an arm, leg, loss of ability to speak, incoordination and so on. Diffuse damage of cerebral tissue, also leads to certain symptoms. For example, in mild diffuse diseases, failure of various functions like attention, concentration, judgment and memory are found (1). Other symptoms include changes in alertness, like stupor and coma. Some symptoms occur in both focal and diffuse damage, like seizures (6).

Going back to the main causes of damage, various diseases states are involved. Examples on conditions that cause focal damage include :

a.  Stroke, caused by arterial occlusion or hemorrhage.
b.  Trauma.
c.  Cerebral abscess.
d.  Tumors.

In addition to focal damage of the brain and spinal cord, some of these conditions cause changes in CSF, by damaging the blood brain barrier (BBB) and then causing different pathological states. For example, inflammation and tissue necrosis (1).

### 1.1.2.1  coma and stupor

Coma is a state of unconsciousness, where the patient cannot be awaken. Stupor is the state in which the patient is unresponsive, but can be aroused by a certain stimulator. The causes of coma are metabolic and infectious (7).

### 1.1.2.2  Intracranial bleeding

When bleeding occurs from a vessel on the surface of the brain, blood will be present between the brain surface and the pia and arachnoid. This state is called subarachnoid hemorrhage. It has many consequents, because blood is an irritating substance when it is out of its vascular channels. That is, leukocytes flow into the CSF by the irritated meninges (8).

### 1.1.2.3  Infections and inflammatory diseases

Microorganisms which can invade the brain and meninges, will cause an inflammation leading to the known disease meningitis (1). Microorganisms are of different kinds, consequently meningitis is of different kinds. The types of invading microorganisms include bacteria, virus, fungi and ameba (9), (10). Each type of meningitis has its own characteristic pattern, concerning leukocytes, glucose and protein concentration in CSF.

The symptoms of meningitis, depend on the severity of the disease. In acute meningitis, signs of headache and stiff neck is observed. Many pathological conditions could be associated with acute meningitis, such as confusion, stupor, coma and cerebral infarction. If the inflammation reaches the cranial nerve roots, signs of deafness and weakness may occur.

In subacute and chronic meningitis, tension hydrocephalus may occur. This state happens when an obstruction in the CSF flow. In adults, signs of impairment of consciousness. The CSF pressure rises, the ventricles enlarge and the choroid plexuses are compressed. In infants and young children, the main signs are head enlargement and inability to look upward. Another pathological condition associated with subacute and chronic meningitis is subdural effusion. The main signs are impaired alertness, refusal to eat, vomiting, immobility and fever (2), (11).

Invasion of the brain by organisms is referred to as encephalitis. CSF findings in this disease are comparable to those in meningitis (9). Abscess formation in the brain (brain abscess) may not cause any CSF changes, even though a severe infection may cause death is present.

### 1.1.2.4  Ischemia and stroke

Ischemia, is an insufficient blood flow to a tissue. Many reasons may cause this situation, such as :

a.  Total stoppage of cerebral blood flow, resulting from the stoppage of the heart.
b.  Blood pressure drops to allow level that the flow becomes insufficient.
c.  Narrowing of a large vessel, such as the carotid artery and so too little blood passes through.

Stroke : When a tissue becomes ischemic, blood vessels surrounding this tissue try to supply the area and overcome the difference. If this does not happen and the tissue remains ischemic, that area dies. This situation is termed a stroke or cerebral infarction (12).

### 1.1.2.5 Epilepsy, seizure disorders

A seizure is a paroxysmal, unregulated burst of electrical firing at some point on the cerebral cortex. The region where the seizure happens, maybe limited within the boundary of the abnormal discharge, or it may generalize to large areas of the cerebral cortex. The former situation is called a focal seizure. The symptoms depend on the site of the tissue undergoing abnormal discharge. Generalized seizures can cause loss of consciousness, violent shaking and loss of urinary continence (12).

The cause of seizures in most instances is not known, but they occur in association with other diseases, like birth injuries, brain tumors, strokes, metabolic abnormalities, head trauma, infections and ingestion of toxins (1).

## 1.1.3 The cerebrospinal fluid (CSF)

The cerebrospinal fluid is a clear colorless fluid that surrounds the brain and spinal cord.

### 1.1.3.1 Formation

CSF is mainly formed by a specialized sponge like structure called the choroid plexuses, in the lateral ventricles. The formation of CSF is based on a combined process of active transport and ultrafiltration (13). Both processes are facilitated by two factors, the thin-walled vessels of the plexuses, that allows passive diffusion of substances from blood plasma and the organelles in choroidal epithelial cells that are responsible for the active transport (14). Another source of CSF, is the blood vessels in the subependymal regions and the pia. To this extent, electrolytes and glucose equilibrate with the CSF at all points in the ventricular and subarachnoid spaces (14). The total volume of CSF in adults is approximately 150 mL, about 20 mL is in the ventricles and about 130 mL in the subarachnoid, space distributed between the subarachnoid cisterns (approximately 60 mL) and the spinal canal (approximately 70 mL). In neonates, the total volume of CSF is approximately 10 to 60 mL (15). The rate of CSF formation is about 500 mL/day (1). As a whole, CSF is renewed four to five times daily (14).

### 1.1.3.2 Circulation

As mentioned before CSF is produced in the lateral ventricles. After its formation it circulates into the third ventricle, then the fourth ventricle. CSF exits from the fourth ventricle, by three small openings or foramina of Lushka and Magendie. Then it circulates upward and downward through the intracranial and spinal subarachnoid spaces. The exit of CSF from the ventricles is facilitated by the arterial pulsations of the choroid plexuses and its downward flow occurs under a decreasing gradient of pressure. The gradient of pressure is highest in the ventricles and gradually decreases along the subarachnoid pathways (14).

### 1.1.3.3 Absorption

In order that CSF is produced several times at a constant volume, it must leave the subarachnoid space. CSF is absorbed through the arachnoid villi and granulations. These are "outpouchings of the arachnoid membrane that penetrate the dura and protrude into the venous system around the brain"(1). They represent a network of folded tubes whose walls separate the subarachnoid space from the venous space. The arachnoid villi and granulations, function as unidirectional valves and so the fluid flows in only one direction. In addition, they are capable of clearing very small particles from CSF, like cellular debris from leukocytes and erythrocytes (16).

### 1.1.3.4 Functions of CSF

The functions of CSF could be summarized as follows:

a. Mechanical function, CSF serves as a water jacket for the brain and spinal cord. This gives a kind of protection from injurious strokes. In addition, it reduces a lot of the brain's weight. It was found that the 1400 gm brain, weighs only 50 gm when weighed in water. So the brain virtually floats in its CSF jacket (14).

b. Since the brain and spinal cord have no lymphatic channels, CSF helps removing metabolic waste from the brain.

c. CSF, along with the extracellular fluid compartment help in maintaining the chemical environment of the brain.

d. There is evidence that CSF transports biologically active compounds which may function as chemical messengers (1).

### 1.1.4  Barriers of the brain, blood and CSF

Studies have shown that the different constituents of CSF are in equilibrium with blood.  The CSF in the ventricles and subarachnoid spaces is also in equilibrium with the intercellular fluid of the brain, spinal cord and olfactory and optic nerves.  These nerves represent two of the twelve nerves that form the cranial nerves.  The olfactory nerve serves the sense of smell, whereas the optic nerve serves the sense of sight [14].

The exclusion of many substances in the blood from CSF and the intercellular fluid of the brain and spinal cord is under the control of barriers that differ from each other in their permeability for the different plasma constituents [14].  These barriers are: the blood-CSF barrier, CSF-brain barrier and blood-brain barrier.

**Blood-CSF Barrier :**The different concentration of many solutes in plasma and CSF are attributed to the Blood-CSF barrier.  The latter is represented by the choroid plexus epithelium and also the endothelium of all capillaries in contact with the CSF [15].

**CSF-Brain Barrier :**This barrier is represented by the pia matter of the CNS which is the innermost covering of the brain and lies directly on it.  Across this barrier the equilibrium between CSF and intercellular fluid of the CNS is maintained [15].

**Blood-Brain Barrier (BBB) :**It is a physiological barrier which separates the brain and CSF from substances borne in the blood so it exists between capillary blood and CSF.  The barrier is represented by the endothelial cells of brain micro vessels which form complex tight junctions that prevent diffusion from blood to brain.  Consequently, both brain and CSF composition is maintained at a level different from that of blood with respect to proteins, ions, and other molecular elements [17].

The blood brain barrier has a great importance in clinical practice :

a.    It determines the entering of antibiotics to the brain and meninges.

b.    It helps maintaining the brains chemical environment, despite any change that might occur in the peripheral blood.

The access of substances to the brain and CSF depend on many factors [1], these include :

a.  Molecular weight: the access is inversely related to size.

b.  Protein binding: protein bound substances do not enter the CNS easily compared with those that are unbound. Examples on protein bound substances are some of the drugs, calcium, magnesium, and metabolites like bilirubin.

c.  Lipid solubility: high lipid soluble substances like carbon monoxide and alcohol easily enter the CNS.

Highly polar substances, like some amino acids, enter slowly and require an active transport mechanism. A well known glycoprotein γ-glutamyl transpeptidase (GGT) (E.C. 2.3.2.2), which is a biochemical marker protein for the endothelial cells of the brain micro vessels, may serve in the amino acid transport across the barrier [18].

## 1.1.5  CSF specimen collection

CSF could be obtained by a. Lumbar puncture (LP). b. Cisternal and cervical subarachnoid space puncture. c. Lateral cerebral ventricle. LP is preferred, but risky in certain cases like when CSF pressure is high [19]. LP is performed at the L3 - L4 or lower to avoid damage to the spinal cord. In infants and young children the cord may extend as low as L3 - L4, so the puncture is performed at the L4 - L5 or lower [15].

## 1.1.6  CSF Composition

The composition of CSF differs from that of serum in some components and is the same for others. In neurological diseases, CSF examination yields valuable information which are important in the diagnosis of the disease [20].

CSF composition includes :

1.  Pressure.
2.  Gross appearance.
3.  Cells and presence of microorganisms.
4.  Protein.
5.  Glucose.
6.  Electrolytes and acid-base measurements.
7.  Enzymes.

### 1.1.6.1 Pressure

The pressure of CSF is maintained by systemic circulatory factors. It is related to blood flow and volume in the arteries and veins of the head. Rise in arterial pressure causes a slight or no increase of pressure at the capillary level in which CSF pressure is in equilibrium with. Hence CSF pressure will not increase. In contrast venous pressure has a direct effect on CSF pressure. In particular jugular and vertebral veins that communicate with the intracranial venous sinuses(15).

Before any fluid is withdrawn, the pressure should be measured. This is done by a water manometer joined to a needle. The latter is placed in either the lumbar subarachnoid space or the cisterna magna. The patient should be horizontal in the lateral reclined position. Normal pressure at this position ranges between 80-200 mmH$_2$O. If the patient is sitting, the pressure will reach 280 mmH$_2$O which represents the level of cisterna magna (19).

Many pathological conditions affect the pressure of the brain. These include tumors, hydrocephaly, brain swelling, clots of blood and increased CSF volume. One of the most common diseases related to pressure, is benign intracranial hypertension. Symptoms of headache, blurred vision and dizziness are observed(14).

### 1.1.6.2 Gross Appearance

Normal CSF is crystal clear and colorless, with viscosity comparable to water(15). It is best to examine CSF in a glass tube and compare it with a tube of water. Both tubes are held against a pure white background. Another way is to look down the tube from above. At least 1 mL of fluid should be observed (1).

Abnormal CSF could be turbid, grossly bloody, xanthochromic and even clear. Turbidity is graded from 0 to +4 depending on the degree of turbidity, for CSF could be faintly cloudy with slight turbidity or clearly turbid. The cause of this appearance maybe the presence of : a. Leukocytes, at least 200 cell/µL. b. erythrocytes, at least 400 cells/µL. c. microorganisms, like bacteria, fungi and amebas (21).

Gross blood which is red cells on microscopic examination, may cause a confusion in the diagnosis. Sometimes a considerable amount of fresh blood is present in the specimen, due to a traumatic puncture (22). This confuses with the pathological bleeding caused by subarachnoid hemorrhage, intracerebral hemorrhage, or trauma (15).

Xanthochromia, refers to a yellow color in CSF. It appears in certain cases:

a.  When blood is mixed with CSF, in this case the yellowing requires from 2-4 hours to occur. In order to distinguish pathological bleeding from traumatic tap the CSF specimen should be centrifuged immediately, for bleeding that is caused by a traumatic tap should not produce xanthochromia after centrifugation.

b.  When protein is present in concentrations greater than 1500 mg/L.

c.  When a patient is jaundiced, that is, when the level of bilirubin in serum rises, as in liver failure. In this case bilirubin may enter the CSF. Xanthochromia from bilirubin requires at least 100-150 mg/L serum bilirubin (23).

### 1.1.6.3 Cells and Presence of Microorganisms

Normal CSF contains no cells or at most, a leukocyte count of 0 to 5 mononuclear cell (lymphocytes and monocytes) (24). The leukocytes could be counted by an ordinary counting chamber. Elevation of leukocyte in CSF indicates a reactive process to bacteria or other infectious agents, blood, chemical substances, or tumors (19).

Bacteriologic cultures are performed and in bacterial meningitis, the most valuable single examination is the Gram's stain of CSF (15). In fungal meningitis the classic examination is India ink preparation, but this ink is positive in only 50% of cases. About 80% could be detected by special stains like mucicarmine and methenamine silver on specimens prepared by Millipore filter or cytocentrifuge. The latter two represent methods for concentrating CSF leukocytes prior to counting (25). Some amebic organism are difficult to detect or identify on Gram's stain or wright's stain. CSF cell counts should be performed as soon as possible, for leukocytes like erythrocytes undergo lysis within about one hour (15). For tuberculous meningitis acid-fast stains are usually performed on CSF. It was found that the more sensitive stain is the fluorescent rhodamine stain (26).

### 1.1.6.4 Protein

The source of proteins in CSF is : **a.** Ultrafiltration of plasma across blood-CSF barrier. **b.** Synthesis within the CNS. Most of the plasma proteins are removed by the process of ultrafiltration. The total protein concentration of lumbar CSF is 15-45 mg /dL (15), (22), (27). While that from cisterns is 10-25 mg/dL and that from the ventricles is 5-15 mg/dL. These levels are much lower than the protein level of serum 600-780 mg/dL (27).

Diseases that affect the permeability of the barrier, generally lead to an increase in total CSF protein. For example brain tumors, bacterial meningitis and trauma. Further fractionation of CSF protein constituents is carried out by immunoelectrophoresis. The main immunoglobulins in CSF are IgG, IgA, and IgM. It also contains trace amounts of IgD and IgE. Among these immunoglobulins, IgG is the most important, especially in some demyelinating diseases such as multiple sclerosis (28 ), (29 ).

**Methods of CSF protein analysis :** Protein determination represents an integral part of CSF analysis. Different methods are used for measuring total protein, these include :

a. **Lowry method :** This chemical method uses Folin phenol reagent and involves two steps. First, protein reacts with copper in an alkaline solution. Then the copper protein complex and any tyrosine and tryptophan present, reduce phosphotungstic-phosphomolybdic acids to a color product. It is a sensitive method giving a good color, with 0.1 mg/mL of protein or less and it requires only 100-200µL of CSF (27), (29).

b. **Biuret method :** In this method protein reacts with copper in alkaline solution. It has a very low color yield in the range of CSF protein concentration. However, Finley and Williams described a more sensitive rate biuret method for CSF proteins. This method also requires 100-200 µL CSF(30).

c. **Trichloroacetic acid:** This method abbreviated (TCA) is a turbidimetric method based on the precipitation of protein followed by measurement of turbidity by a spectrophotometer. It requires 500 µL CSF (27).

d. **Sulfosalicylic acid:** A turbidimetric method which uses 3% sulfosalicylic acid and 7% sodium sulfate. Protein is precipitated as a fine white precipitate and the resulting turbidity is measured, it requires 500 µL CSF (27).

e. **Comassie brilliant blue dye binding :** This chemical spectrophotometric method abbreviated (CBB), depends on the binding of comassie brilliant blue G-250 to protein. The color of the protein-dye complex changes from brownish orange to intense blue (31). Modifications of this method have been made in order to decrease variability of the color yield with different proteins(32).

f. **Silver-binding assay :** A recent method developed by Krystal G., *et al.* based on the capacity of glutaraldehyde-treated protein to bind silver. It is favorably compared with (CBB) being approximately 100 times more sensitive. Only 0.5 µL of CSF is required for an accurate protein estimation. So it is very useful in cases where CSF specimens are available in very limited amounts, like specimens taken from children (33).

### 1.1.6.5 Glucose

Glucose enters CSF from plasma by at least two mechanisms, active transport and passive diffusion. CSF glucose is normally in the range 45-80 mg/dL (15), (19). This level represents about two thirds that in blood (80-110 mg/dL). Accurate evaluations of CSF glucose require a relative constant level of plasma glucose. Determination of CSF glucose helps differentiating bacterial from viral meningitis, because the glucose level decreases in bacterial and tuberculous meningitis, whereas in viral disease the level is generally normal (15).

CSF glucose level is considered to be decreased, when it falls bellow 40mg/dL in a fasting patient with normal plasma glucose. Generally, the decrease maybe attributed to impairment of active transport ; to increased utilization of glucose by the CNS tissue, leukocytes, erythrocytes and microorganisms; or to hypoglycemia. In bacterial meningitis, the decrease is believed to be a result of two mechanisms, impaired transport of glucose from plasma to CSF and increased utilization of glucose by CNS tissue, leukocytes and microorganisms. Elevated CSF glucose, absolute or relative to plasma glucose is evidence of hyperglycemia (15).

### 1.1.6.6 Electrolytes and acid-base balance

Electrolytes in CSF are mainly sodium, potassium, chloride, calcium, and magnesium. Values of CSF sodium is nearly the same as in serum. Chloride and magnesium are higher than in serum. While CSF potassium is lower than that of serum. However, a neurological disease does not alter the CSF concentration of these constituents in any characteristic way (19).

Acid-base balance in CSF is of interest in relation to metabolic acidosis and alkalosis. The normal pH of CSF is about 7.31, this is lower than the physiological pH of blood (7.4). The pH of CSF is under precise regulation, in a way that even in the case of chronic metabolic acidosis, it tends to remain unchanged. But when there is a rapid change in blood pH whether in acidosis or alkalosis, the CSF pH changes in parallel (19).

### 1.1.6.7 Enzymes

Many enzymes have been measured in CSF, lactate dehydrogenase (LDH), isocitrate dehydrogenase (ICD), creatine kinase (CK), adenylate kinase (ADK), aldolase (ALD), enolase (ENO), aspartate amino transferase (AST). These enzymes have been found to rise under certain conditions of neurologic disease. However, only lactate dehydrogenase appears to be of diagnostic value especially in bacterial meningitis (34).

## 1.2 Lactate dehydrogenase (LDH)

Lactate dehydrogenase, is one of the nonplasma specific enzymes. The concentration of such enzymes in tissue is very high, compared with that in plasma(35). According to the IUB, LDH belongs to the first class ; the oxido reductases. The classification number of LDH is 1.1.1.27 this is given according to the transferred and accepting groups involved in the reaction. However, LDH catalyzes the reduction of pyruvate ; the last product in the glycolysis. By the use of an NADH molecule, pyruvate is reduced to the L isomer of lactate, while NADH is oxidized to $NAD^+$. This reaction is called lactate fermentation. It is freely reversible, with a $\Delta G$ of -25.1 KJ/mol, this implies that the overall equilibrium of the reaction strongly favors lactate formation (36), (37) :

$$\text{(Pyruvate)} \quad CH_3\text{-}\overset{O}{\overset{\|}{C}}\text{-}COO \quad \xrightarrow{\text{NADH} \quad \text{NAD}^+} \quad CH_3\text{-}\overset{OH}{\overset{|}{C}H}\text{-}COO \quad \text{(Lactate)}$$

### 1.2.1 The functional role of lactate dehydrogenase

The conversion of pyruvate to lactate, happens under anaerobic conditions, as in actively contracting muscle (38). Otherwise, pyruvate faces another fate. When oxygen is present, pyruvate is oxidized to yield acetyl coenzyme A which is then completely oxidized to $CO_2$ by the citric acid cycle (37). LDH is an important enzyme in both glycolysis and gluconeogenesis (39).

The importance of the reaction catalyzed by LDH, particularly the oxidation of NADH to $NAD^+$, arises from the fact that $NAD^+$ is necessary for the continuity of glycolysis in active skeletal muscle and erythrocytes. Hence, the production of energy as ATP (40). $NAD^+$ is the acceptor of hydrogen in the glyceraldelyde-3-phosphate dehydrogenase reaction, which leads to the formation of 1,3-bisphosphoglycerate. This reaction represents the first of the two energy-conserving reaction of glycolysis that eventually lead to the formation of ATP(41).

Lactate and pyruvate diffuse out of the skeletal muscle into the blood and then carried to the liver. This diffusion is facilitated by the high permeability of the plasma membrane, in most cells, to lactate and pyruvate. Because of the high NADH/NAD ratio in contracting muscle lactate is carried to the liver more than pyruvate. This reaction is favored by the low NAD/NADH ratio in the cytosol of liver cells (39). Then pyruvate is converted into glucose through the gluconeogenic pathway. Glucose is then passed through the blood into the skeletal muscle and glycolysis starts again. These reactions together are called the cori cycle (42), (43). cycle is summarized in figure (1.3) :

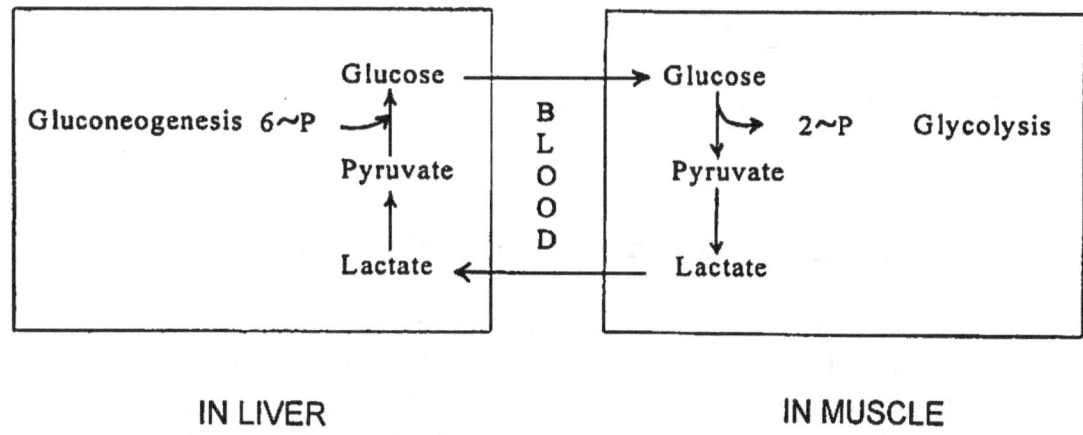

IN LIVER                                        IN MUSCLE

**Figure (1.3).** The cori cycle; the cycle that includes the conversion of glucose to  lactate in muscle, and lactate to glucose in liver.

## 1.2.2 Mechanism of lactate dehydrogenase-substrate-NAD binding

Many studies on the LDH substrate-NAD complex, have been made to understand the way in which the substrate and coenzyme bind to the enzyme. X-ray studies revealed that lactate is bound to the enzyme by its carboxyl group.  This group combines with an arginine in the enzyme, in a way that the hydrogen of the lactate CH group is transferred to the $C_4$ of the adjacent nicotine amide ring and will cause the loss of its charge. At the same time, the hydrogen of the lactate OH group is transferred as a proton to a neighboring histidine group of the enzyme (figure 1.4) (44).

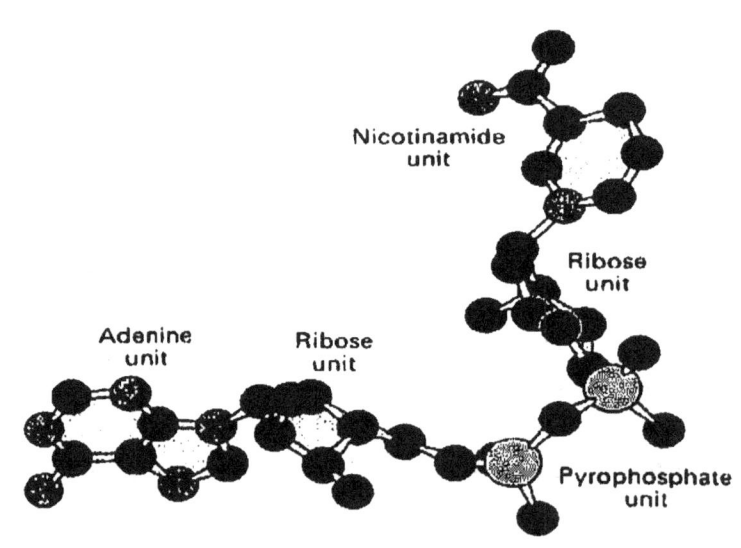

**Figure (1.4).** Diagrammatic representation of lactate dehydrogenase-substrate-NAD binding.

## 1.2.3 Lactate dehydrogenase - NAD binding site

The NAD moecule is composed of four units; the nicotine amide unit, the ribose unit, the pyrophosphate unit and the adenine unit (figure 1.5) . Two of these units are involved in binding; the nicotine amide and adinine unit.The nicotine amide unit binds in a way that the reactive side of the ring is in a polar environment, while the other side is in contact with hydrophobic residues of the enzyme. The adenine part of NAD is bound in a hydrophobic region. As shown in figure (1.5), the bound NAD has an extended conformation (39).

**Figure (1.5).** Model of NAD found in the complex of NAD and lactate dehydrogenase.

The NAD binding region in lactate dehydrogenase, is made up of four helices and a sheet of six parallel strands (figure 1.6). Both nicotineamide binding half and adenine binding half, are structurally the same (39).

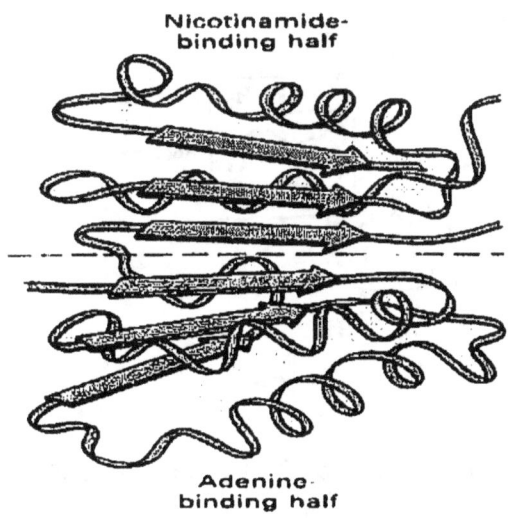

Figure (1.6). Schematic diagram of the NAD-binding region in lactate dehydrogenase.

## 1.2.4  Stereochemical specificity of the nicotineamide ring

The coenzyme NAD, undergoes reversible reduction to NADH. The part of the molecule involved in this reduction is the nicotineamide ring. The source of hydrogen atoms is the substrate molecule lactate, which gives up two hydrogen atoms. One of them, is a hydride ion ( :H ). NAD accepts this hdride ion and converts to NADH. The second hydrogen ( H ) is released to the aqueous solvent. The following reaction illustrates the reduction :

$$NAD^+ + 2e^- + 2H^+ \longrightarrow NADH + H^+$$

The total concentration of NAD and NADH in most tissues is about $10^{-5}$ M.

The nicotineamide ring has two sides. So, the transfer of the hydride ion could be to either sides. The transfer to the front side is called A type, while that to the back side of the ring is called B type (figure 1.7) :

**Figure (1.6).** The transfer of the hydride ion to the front side (A type) and back side (B type) of the nicotineamide ring.

The question is, to which side of the ring will the hydride ion transfer to. Studies using isotopically labeled substrates, revealed that the enzyme catalyzes only one type of transfer and not both. LDH was found to catalyze the A type of transfer (45) (46).

## 1.2.5 Lactate dehydrogenase Isoenzymes

LDH was one of the first enzymes found to have isoenzymes. It is present in vertebrate tissues as at least five isoenzymes. These isoenzymes consist of four polypeptide chains or subunits each with a molecular weight of 35 KD, so LDH is a 140 KD tetramer (37). The subunits are of two types encoded by similar genes; the H type (Heart) and the M type (Muscle). These subunits associate to form five types of tetramers :$H_4$, $H_3M_1$, $H_2M_2$, $H_1M_3$, $M_4$. In the heart, the predominant isoenzyme contains four H type subunits. Whereas, in skeletal muscle and liver the predominant isoenzyme contains four M type subunits (39).

Depending on the different properties of each isoenzyme, many techniques have been used to separate these isoenzymes. The more common technique is electrophoresis, which is based on the different mobility of each isoenzyme in an electric field (47). The five LDH isoenzymes were given numbers from 1 to 5, according to their mobility toward the anode. The fastest isoenzyme is designated $LDH_1$, the slowest is $LDH_5$. So in descending order, the five isoenzymes are : $LDH_1$, $LDH_2$, $LDH_3$, $LDH_4$, $LDH_5$. This designation corresponds to : $H_4$, $H_3M_1$, $H_2M_2$, $H_1M_3$, $M_4$, respectively (48).

### 1.2.5.1 Isoenzyme Distribution

The proportion of LDH isoenzyme in each tissue varies according to the metabolic requirements of that tissue (20). The following table shows the distribution of LDH isoenzymes in various organs of the human body described as percentage of activity (35).

Table (1.1) : The distribution of lactate dehydrogenase isoenzymes in different organs (35).

| Organ | LDH isoenzyme distribution ( % of activity ) | | | | |
|---|---|---|---|---|---|
| | $H_4$ | $H_3M_1$ | $H_2M_2$ | $H_1M_3$ | $M_4$ |
| Heart | 60 | 30 | 5 | 3 | 2 |
| Kidney | 28 | 34 | 21 | 11 | 6 |
| Cerebrum | 28 | 32 | 19 | 16 | 5 |
| Liver | 0.2 | 0.8 | 1 | 4 | 94 |
| Skeletal muscle | 3 | 4 | 8 | 9 | 76 |
| Skin | 0 | 0 | 4 | 17 | 79 |
| Lung | 10 | 18 | 28 | 23 | 21 |
| Spleen | 5 | 15 | 31 | 31 | 18 |

Many factors affect the distribution of these isoenzymes. The development of tissues from their embryonic to adult form represents one of the main factors (37). Changing biological conditions and pathological processes (49), such as ischemia, atherosclerosis, or cancer also affect this distribution (50), (51), (52). Wilhelm showed that among the total LDH activity in normal adult aortic tissue, $LDH_3$

predominates. In atherosclerotic aortic tissue, $LDH_5$ predominates (51). Likewise, myocardial maximal LDH activity is present in $LDH_1$, but during the progress of ischemic heart disease, the activity shifts to $LDH_3$ (50). In lymphoid malignancies, the main LDH isoenzyme in serum are $LDH_2$, $LDH_3$, and $LDH_4$. The predominance of these isoenzyme results from malignant proliferation in which the number of lymphoid cells is increased (49). In many tumors, $LDH_5$ is found to be the predominant isoenzyme such as in colonic cancer tissue (35), (53).

Many investigators have reported unusual electrophoretic patterns and additional bands of serum LDH. Kalpaxis D., found an abnormal isoenzyme in serum, from a patient with hepatocellular carcinoma. This extra isoenzyme named $LDH_1$ ex, migrated faster than $LDH_1$ on agarose electrophoresis. $LDH_1$ ex was isolated and some of its physical and biochemical characteristics were studied (54). Giannoulaki E., also reported the presence of $LDH_1$ ex in different malignant diseases (55).

Lubin and Bhagavan reported an additional LDH isoenzyme designated $LDH_6$, which migrated slower than $LDH_5$. This isoenzyme was found in serum, in some cases of severe hypotension or impaired ventilation. The autopsy material in these cases was studied, and results suggested that the source of this isoenzyme is liver (56), (57). These results were confirmed by other studies, but small amounts were also detected in kidney and spleen. The significant characteristic of this isoenzyme, is that it is lactate independent and its activity enhances when ethanol is added to the substrate mixture. From these results, $LDH_6$ was not recommended a true isoenzyme of LDH. But it was suggested to be a useful serum marker for severe liver injury and as a tissue marker at autopsy for extended postmortem liver damage (58).

Otsu, found an abnormal LDH isoenzyme in serum and tumor tissue of a patient with neuroblastoma, which located between $LDH_2$ and $LDH_3$ on agarose gel electrophoresis (59) . The mobility of this isoenzyme is the same as the extra isoenzyme found in normal human erythrocytes, designated LDH "Y" by Zail and Van den Hoek (60). This isoenzyme is also found in serum in cases of hemolytic anemia. The investigators believe that the abnormal isoenzyme originates from the tumor, so it was suggested that this abnormal isoenzyme could be useful as a marker enzyme for neurogenic tumors (59).

A human trophoblastic isoenzyme designated LDH Z, was found to be present in the first-trimester placenta, hydatidiform mole and cell lines derived from choriocarcinoma as well as an autopsy sample of choriocarcinoma metastasized to the liver. This isoenzyme migrated slower than $LDH_2$ on starch electrophoresis. It was found to be lactate dependent in the presence of NAD and pyruvate dependent in the presence of NADH (61). LDH Z, was earlier observed in the same three types of material (first-trimester placenta, hydatidiform mole and choriocarcinoma cell line) (62), (63). The electrophoretic mobility of this isoenzyme in all three types was found to be the same. Results have shown that LDH Z is a useful markers for revealing the origin of the tumor (61).

Another unusual LDH isoenzyme designated LDHk, was found in human cancer and characterized by an activity level greater than that of other LDH isoenzymes usually associated with human cancer. Before this isoenzyme was found in human cancer, it was identified in cells infected by kurtsun murine virus (KiMSV). The unusual features of this isoenzyme, is that it is inactive when assayed in the presence of oxygen, strongly inhibited by GTP and related compounds and more basic than the other isoenzymes of LDH with a P above 8.5 . Nondenaturating gel electrophoresis was used for the detection of LDHk. One of the explanations given about the presence of LDHk in a majority of human cancer, is that the abnormal expression of the human gene, that caused cancer, is similar to that of the rat genes of KiMSV directly created these cancers (64).

Abnormal patterns of LDH isoenzymes were found in patients with pericarditis and myocardial hypertrophy. The abnormality was attributed to the formation of a complex between LDH and immunoglobulins, such as IgA, IgG, IgM (65), (66), (67). The formation of some complexes results in a decrease in LDH activity. Sudo, observed a low LDH activity and an abnormal electrophoretic pattern of the isoenzymes on a cellogel membrane. This result was attributed to the formation of a complex between LDH and IgG, and not IgA or IgM (68). Wickus and Smith, found a loss of LDH activity when a LDH-IgG complex was formed, but this complex did not show the abnormal isoenzyme pattern obtained by Sudo(66).

### 1.2.5.2 Methods OF LDH Isoenzyme Analysis

Many physical and catalytic differences among LDH isoenzymes have been exploited in order to measure the concentration of these isoenzymes. Different techniques have been used for such purpose :

a.  Physical methods : These include ion-exchange chromatography, heat stability and electrophoresis (36), (69). The most common electrophoretic methods used, are polyacrylamide gel electrophoresis (PAGE) and agarose gel electrophoresis (AGE) (61), (69). Other electrophoretic methods include electrophoresis on cellulose acetate plates, disc electrophoresis and ultrathin-layer zone electrophoresis (70), (71), (72).

b.  Chemical methods : These include substrate-product relationship, coenzyme affinity and catalytic inhibition (36), (35).

c.  Immunological methods : These methods in general, have become an important and wide spread technique. They include radioimmuno assay (RIA), immuno radiometric assay (IRMA) and immunoinhibition techniques (35). These techniques are based on the different amino acid sequence and thus the different immunological characteristics of subunit polypeptides. For LDH, the more frequent technique is the immunoinhibition or immunoprecipitation assays for LDH₁ (71). The method depends on using antibodies against the M subunit (anti-M), in order to determine LDH₁. The antibodies will cause all the M bearing subunits to be removed from the solution leaving the subunits of H type only, that is LDH₁. Then, LDH₁ is assayed by an ordinary LDH procedure (71). This method is considered to be more sensitive than the usual electrophoretic analysis for the detection of early elevations of LDH₁ after acute myocardial infarction (AMI) (73).

### 1.2.5.3 Methods of LDH assay

Different methods have been applied to the assay of LDH, spectrophotometric, colorimetric and fluorometric. But the most frequently used method is the spectrophotometric. This method depends on measuring the interconversion of the coenzyme NAD and NADH at 340 nm (74), (75). The reaction is either followed in the lactate to pyruvate (L→P) or pyruvate to lactate (P→L) direction (76). End-point and kinetic modes of analysis could be used for both directions (47). Older methods are based on using an indicator dye to determine the amount of reactants consumed or formed. Dinitrophenyl hydrazine and tetrazolium dyes are both used as indicator dyes. Dinitrophenyl hydrazine forms a colored product with pyruvate, while tetrazolium oxidizes NADH and is reduced to form a colored tetrazolium dye (77).

The common methods which are routinely used for LDH assay are wroblewiski and LaDeu, Wacker method, Scandinavian method and Henry method (78), (79), (47). These methods differ from each other in the direction of the reaction followed, type of buffer used, temperature and the concentration of substrates.

### 1.2.6 Lactate dehydrogenase kinetic properties

The accuracy of enzyme activity measurements, depends on the linearity of the reaction. The linearity of the LDH catalyzed reaction, mainly depends on the direction of reaction (pyruvate to lactate or lactate to pyruvate). Many studies were made and for both directions, different results were obtained. A study on the LDH assay for the pyruvate to lactate direction, showed that the linearity depends greatly on the initial concentration of NADH and that the reaction was linear up to 100 mol/L NADH. Higher concentrations led to a nonlinear reaction. This result was attributed to the substrate inhibition caused by high NADH concentrations. High concentrations of pyruvate, was also found to cause substrate inhibition. However, another study on the pyruvate to lactate direction revealed less substrate inhibition for NADH (74),(80).

Nonlinearity is also observed in assays performed in phosphate buffers and in the lactate to pyruvate direction. This has been attributed to the formation of a ternary complex between NAD, pyruvate and phosphate. However, in a recent study investigation of the possible causes of nonlinearity during LDH assays in Tris buffers and in the lactate to pyruvate direction, indicated that the inhibition of LDH activity by pyruvate may be the main cause of nonlinearity. This result was mainly observed in serum exhibiting abnormally high LDH activity. Most of the LDH activity in serum was inhibited by 5 mM pyruvate. By contrast, most of the LDH isoenzyme activities were inhibited partially at 0.5 mM pyruvate, $LDH_1$ being the most and $LDH_4$ the least affected. In assays of serum with high LDH and pyruvate concentration using Tris buffer at pH 9.3, the addition of a bacterial pyruvate oxidase (PO) was found to remove the accumulation of pyruvate, hence its inhibitory effect was overcame. So it was possible to assay serum exhibiting high levels of LDH activity without being affected by pyruvate (81).

A study on the lactate to pyruvate direction, showed a wide range of linearity. The study depends on a regression-kinetic procedure. This differs from other kinetic procedures which depend on changing dilution factors and measurement intervals, for obtaining a more sensitive and linear range (82). However, disagreement about following the reaction in the lactate to pyruvate or pyruvate to lactate direction remains, but since there is no general accepted method for LDH assay, following the reaction in both directions is adequate for routine work(47).

Finding out optimum conditions for the enzyme catalyzed reaction, leads to more accurate kinetic parameters. A study on serum LDH in the pyruvate to lactate direction at pH 8.7 and at (25, 30, 37)°C, gave a different optimum concentration for both NAD and lactate at each temperature. The highest optimum concentration was found at 37°C (83). For the pyruvate to lactate direction and at pH 7.0, NADH had the same optimum concentration at (25 and 30) °C and a high optimum concentration at 37°C. For pyruvate the same highest optimum concentration was observed at (30 and 37)°C (76). Optimum substrate concentration for LDH isoenzymes, differ from one isoenzyme to another. For example, $LDH_1$ and $LDH_5$ from serum of patients with kalazar, gave an optimum pyruvate concentration of 5 mM and 18 mM, respectively. For NADH, the optimum concentration for $LDH_1$ and $LDH_5$ was 2 mM and 6.2 mM, respectively (84).

The most important kinetic parameter in enzyme studies, is the Michaelis-Menton constant ($K_m$). This constant reflects the affinity of the enzyme to the substrate. $K_m$ values for lactate, for $LDH_1$, $LDH_2$ and $LDH_3$ isolated from different human tissues were found to range between $(1 \times 10^{-5} - 2 \times 10^{-3})M$ and $(4.2 \times 10^{-5} - 5.6 \times 10^{-5})M$. $K_m$ for pyruvate for $LDH_1$ isolated from human heart, was found to have a value of $(1.18 \times 10^{-4} \mp 0.12)M$. The $K_m$ value for pyruvate for LDH isoenzymes isolated from animal tissue was also studied. From rabbit muscle and erythrocytes, the $K_m$ value for pyruvate for $LDH_1$ was found to be $(6.7 \times 10^{-5})M$. This value is less than $k_m$ for $LDH_5$ isolated from the same source $(3.5 \times 10^{-4})M$. $K_m$ for pyruvate for $LDH_5$ isolated from different animal sources, showed a higher value than that of $LDH_1$ about 2-10 times (84).

The temperature of the reaction is of great importance, because the optimum conditions and enzyme response vary according to temperature. LDH isoenzymes differ in their response to temperature. $LDH_1$ is the most heat stable isoenzyme, it was found that when LDH is heated at 65°C for 1.5 hour all isoenzymes lose their activity except for $LDH_1$ (78). The thermal stability of serum LDH isoenzymes from a with hepatocellular carcinoma was tested, and at different temperatures from 10 - 60°C. Up to 20°C, all isoenzymes retained their activity, but at 60°C the most stable isoenzyme was $LDH_1$. Whereas, the most labile isoenzyme was $LDH_4$, its activity was completely lost at this temperature. The order of the LDH isoenzymes stability at 60°C was : $LDH_1 > LDH_2 > LDH_3 > LDH_5 > LDH_4$ (54).

The pH of the enzyme catalyzed reaction is very important, because it has an effect on the dissociation of the substrate and certain amino acids in the active center or elsewhere in the enzyme molecule. It also has an effect on the three dimensional structure of the protein and therefore on the enzyme activity (85). The pH optimum for the LDH catalyzed reaction in a certain differs than that in the reverse reaction. In general, for the lactate to pyruvate direction and at 37°C, the pH optimum is 8.7. Whereas, for the pyruvate to lactate direction pH optimum is 7.0(47).

It was found that high concentrations of pyruvate and lactate inhibits LDH activity. $LDH_1$ is more affected by high substrate concentration than $LDH_5$. Product inhibition is also observed and for both the lactate to pyruvate and pyruvate to lactate direction (47). Product inhibition also affects $LDH_1$ more than $LDH_5$. It was found that a lactate concentration of 21.9 mM, inhibits 55% of $LDH_1$ activity and 14% of $LDH_5$ activity (86). LDH activity is also affected by other substances rather than the substrates and products. Both urea and oxalate have an inhibitory effect on LDH activity. A concentration of 2 M urea is enough to inhibit LDH. $LDH_5$ is more inhibited than $LDH_1$ at such concentration. In contrast, 0.2 mM of oxalate inhibits $LDH_1$ more than $LDH_5$ (87). Other substances like nucleotides, citric acid and oxaloacetic acid, were also found to have an inhibitory effect on LDH (87).

### 1.2.7 Lactate dehydrogenase in the cerebrospinal fluid

LDH is a normal constituent of brain tissue and the latter is rich in LDH. Normal CSF has a low level of LDH activity, for it was found that CSF LDH activity is approximately 10% of serum (1), (88). The range of normal serum LDH activity varies depending on the direction of the enzyme reaction and the type of method used. For the pyruvate to lactate direction and at 30°C, LDH activity is in the range of 90 - 310 U/L, while that for CSF, at the same conditions, is 7 - 30 U/L (89), (78). LDH in CSF, comes from different sources :

a.      From blood by diffusion across the blood-CSF barrier.

b.      From CNS by diffusion across the brain - CSF barrier and because the brain is rich in LDH, any damage of CNS tissue will increase the level of CSF LDH.

c.      Cellular elements in CSF, such as leukocytes, bacteria and tumor cells (15).

An increase in LDH activity has been reported in many different diseases. Feldman, found in his study that about 90% of patients with bacterial meningitis had an increase in LDH activity, versus 10 % of those with viral meningitis (90). Jain et al. reported a significant increase in LDH activity in pyogenic meningitis compared with that of tuberculous meningitis or encephalitis. They also found a direct relationship between the enzyme activity and CSF protein content as well as total cell count (91). A rise in LDH activity is also found in cases of subarachnoid hemorrhage and metastatic carcinoma involving the CNS (92), (93). It is worth mentioning that many tumors utilize glucose through the anaerobic pathway and as a result, a considerable amount of lactate is produced. This phenomenon led many investigators to study LDH in patients harboring tumors (70).

### 1.2.7.1  Lactate dehydrogenase isoenzymes in the cerebrospinal fluid

As a source of energy, the brain tissue mainly depends on aerobic metabolism since the brain of adult mammals normally uses only glucose as fuel (94). To this extent, normal brain has a predominance of the aerobically active isoenzymes of LDH which are $LDH_1$ and $LDH_2$ (93).

LDH isoenzyme analysis in CSF has improved the specificity of LDH measurement in CSF (15). Recent studies show that LDH isoenzyme analysis is helpful in differential diagnosis of various CNS disorders. Chatterley evaluated the diagnostic value of LDH by studying 93 CSF specimens taken from patients with diseases of four categories; tumors, infections, hemorrhages and others (34). In metastatic tumors, the predominant isoenzyme was $LDH_5$, while primary brain tumors showed an increase in all fractions. Concerning infections, an increase in $LDH_4$ and $LDH_5$ was found in bacterial meningitis while viral encephalitis revealed an increase in $LDH_1$, $LDH_2$ and $LDH_3$. Subdural and subarachnoid hemorrhages showed an increase in all isoenzymes (34).

Some of these results came in agreement with other studies (93). Concerning the predominance of $LDH_5$ and $LDH_4$ in malignancy, the investigators agree that during the course of malignant transformation, cells obtain a large part of their energy from anaerobic processes. Consequently, there is an increase in the anaerobic isoenzymes $LDH_4$ and $LDH_5$. This phenomenon has been noted in a variety of cancer tissues including the brain and it appears that the more malignant the tissue the greater the $LDH_5$ to $LDH_1$ ratio (95), (96).

LDH in CSF has shown to be useful in many neurological disorders, but still its sensitivity awaits further improvement (34).

# Aim of work

According to the current situation of lactate dehydrogenase, the relevant studies on this enzyme in the cerebrospinal fluid indicate the importance of this enzyme in this biological fluid. But most of them give special reference to activity measurements and lack some of the important biochemical aspects. As such the intention to study the following lines was pursed :

a. Protein and activity measurements of this enzyme in some of the diseases related to the central nervous system, such as bacterial meningitis and hydrocephaly.

b. Separation of this enzyme from CSF and purification of its isoenzymes $LDH_2$ and $LDH_5$ by column chromatography methods (gel filtration and ion exchange).

c. Kinetic and thermodynamic studies on the isolated $LDH_2$.

d. Spectroscopic studies on $LDH_2$ and $LDH_5$, including binding studies on LDH5 with NADH.

# CHAPTER TWO

## EXPERIMENTAL

## 2.1 Equipment, materials, chromatography and CSF samples

### 2.1.1 Equipment used :

a.  Spectrophotometer (LKB ultraspec 4050).
b.  Electrophoresis system (LKB 2117 multiphor) with a power supply.
c.  Cold room (2201 combicoldrac II) with its accessories.

### 2.1.2 Materials used

### 2.1.2.1 General chemical reagents

All common laboratory chemical reagents were of Analar grade or the equivalent, unless otherwise specified and where obtained from the following companies :

a.  BDH company :

Tris (hydroxymethyl) aminomethane.
Nicotineamide adenine dinucleotide (reduced).
Nicotineamide adenine dinucleotide.
Sodium pyruvate.
Lithium lactate.
Bovine serum albumin.
Acrylamide.
Ammonium persulphate.
N, N methylenebis acrylamide (BIS).
Nitro blue tetrazolium (NBT).
Phenazine methosulphate (PMS).
Temed.

b.  Merk company :

Folin-ciocalteau reagent.

c.  Pharmacia Fine chemicals :

Sephadex G-150.
DEAE-Sephadex A-50.

### 2.1.2.2  Buffer systems

All buffer solutions were prepared by dissolving the appropriate amount of salt in deionized distilled water and the required pH was adjusted by the addition of HCl in the case of Tris-HCl.

### 2.1.3  Chromatography

Chromatography material were supplied as dry powder.  The dry powder of Sephadex G-150, DEAE-Sephadex A-50 was allowed to swell in an excess of de-areated buffer, according to the manufacturer's instructions.  The process of swelling can be accelerated by using a boiling water bath, which is also serves to de-areate the buffer.  Fine particles were removed by decantation before packing the column.  The packing of a column is a critical stage in any chromatographic process.  The matrix suspension should be adjusted to a thick slurry, usually about 75% of settled matrix.  The suspension must be carefully mixed before pouring into a vertical column containing eluant buffer.  After the gel has settled, two bed volumes are used at a slightly higher flow rate.  Constant pumping speeds were achieved using 1200 varioperpex peri-staltic pump (LKB).  All fractions were collected  at 4° C on LKB 7000 ultra rac fraction collector.

### 2.1.4  Cerebrospinal fluid samples

CSF was obtained from patients admitted to the Neurosurgical Hospital/ Baghdad. Patients of different ages  suffered from a variety of neurological diseases, mostly bacterial meningitis and hydrocephaly.  The fluid was withdrawn by lumbar puncture, using a spinal needle (No. 20).  The patient lies on a hard bench, taking the lateral reclined position and the needle is gently placed above or beneath the forth lumbar vertebra.  The amount of withdrawn CSF is not fixed , but usually in the range of (1-5) mL.  Freshly collected specimens were stored at 4°C. Turbid specimens were centrifuged at 3000 xg for 10 minutes before storage.

## 2.2  Lactate dehydrogenase studies

### 2.2.1  CSF protein measurements

The protein concentration of CSF samples, was determined by the Lowry method, using bovine serum albumine (BSA) as the standard protein [97].

**Materials :**

1.     Solution A (alkaline sodium carbonate solution) :  2 gm of sodium carbonate ($Na_2CO_3$) is dissolved in 100 mL of 0.1 N sodium hydroxide (NaOH).

2.     Solution B (copper sulphate-sodium potassium tartrate) :  0.5 gm of copper sulphate ($CuSO_4 . 5H_2O$) is dissolved in 100 mL distilled water.  From this solution, 10 mL is taken and 0.1 gm of sodium potassium tartrate is added.  This is prepared on day of use.

3.     Alkaline solution :  This is prepared on day of use by mixing 50 mL of solution A and 1 mL of solution B.

4.     Folin-Ciocalteau reagent : This reagent is a solution of sodium tungstate and sodium molybdate in phosphoric acid and hydrochloric acid.  The reagent is diluted with an equal amount of distilled water on the day of use.

5.     Standard protein :  The standard protein (BSA), is prepared as follows :

   a.     A stock solution of 1000 µg/mL is prepared by dissolving 100 mg BSA in 100 mL distilled water.

   b.     From the stock solution, the following concentrations are prepared by serial dilution with distilled water : (200, 150, 100, 50, 25) µg/mL.

6.     Sample preparation : 0.1 mL of CSF is diluted by an appropriate     amount of distilled water.

**Procedure :**

1.     2.5 mL of alkaline solution is added to 0.5 mL of standard protein or diluted CSF sample and mixed thoroughly.  Then the mixture is left for 10 minutes.

2. 0.25 mL of Folin-Ciacalteau reagent is added and mixed immediately and rapidly. This mixture is left for 30 minutes then the absorbance is read at 600nm.

3. The standard curve was obtained by plotting the absorbance against the corresponding concentration of standard protein and used to determine the unknown protein concentration of the CSF sample (figure 2.1).

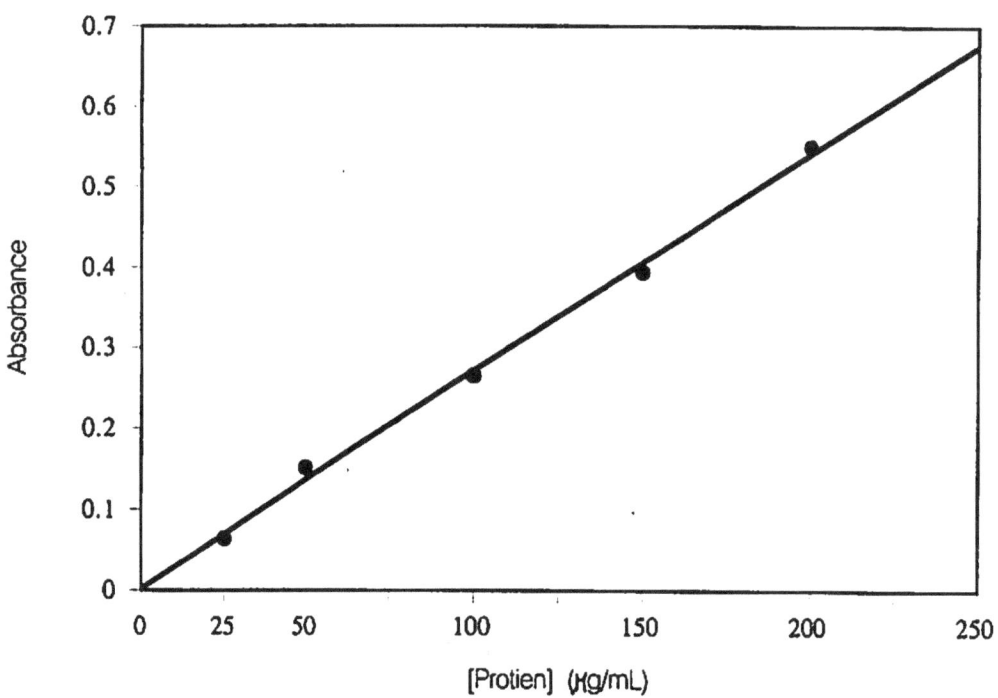

Figure (2.1). The standard curve for protein dtermination by the Lowry method.

**Immunological measurements** : The protein fractions measured in this study were albumin, and the immunoglobulins IgG and IgA. These proteins were measured by radial immuno diffusion (98).

## 2.2.2 Activity measurements of LDH in CSF

The activity of LDH was measured according to the Scandinavian method (Scandinavian committee on enzymes recommended method). In this method, the reaction is followed by measuring the rate of NADH consumption at 340 nm.

**Materials :**

1.   Working Tris buffer (0.056 M) : 6.8 gm of Tris (hydroxymethyl) aminomethane is dissolved in approximately 800 mL of distilled water. The pH is adjusted to 7.4 with 1 M HCl and the volume of solution is brought to 1 L by distilled. This solution is stable for at least 6 weeks at 4°C.

2.   Tris-NADH reagent : 13 mg of NADH is dissolved in 90 mL of working Tris buffer. The absorbance is measured and brought to an absorbance of 1.0 (161 m.) by dilution with the same buffer. This solution is stable for 72 hours at 4°C.

3.   Pyruvate working solution (13.5 mM) : 149 mg of sodium pyruvate is dissolved in 100 mL of distilled. This solution is stable for 20 days at 4°C.

**Procedure :**

1.   1 mL of both Tris-NADH reagent and working Tris buffer is placed in a 3 mL cuvette.

2.   0.1 mL of CSF is added and the solution is mixed thoroughly, then incubated for 10 minutes at 30°C.

3.   The reaction is initiated by the addition of 0.1 mL of working pyruvate solution. After mixing, the cuvette is rapidly inserted into the spectrophotometer and the change in absorbance at 340 nm is immediately.

### 2.2.2.1 Initial rate determination

For initial rate measurement, the change in NADH concentration was followed throughout a period of time. The change in NADH concentration was plotted versus time and the slope was determined. The slope represents the initial rate.

## 2.2.2.2  Specific activity determination

The specific activity was determined according to the following equation :

$$\text{Specific activity} = \frac{\text{Activity}}{\text{Protein amount}}$$

## 2.2.3  Purification of lactate dehydrogenase isoenzymes in CSF

### 2.2.3.1 Purification by gel filtration and ion exchange chromatography (pH gradient)

LDH was purified by two methods of column chromatography.  First gel filtration on Sephadex G-150, subsequently ion exchange on DEAE-Sephadex A-50 for further separation and as follows :

1. Activity measurement : The activity of CSF LDH was measured as described in section (2.2.2).

2. Protein determination : The protein concentration of the CSF sample was determined as described in section (2.2.1).

3. Gel filtration :

**Materials :**

a. Sephadex G-150.
b. Tris (hydroxymethyl) aminomethane (0.056 M, pH 7.4).
c. Blue dextran 2000.

**Procedure :**

a. Gel preparation and column packing :

Gel filtration was carried out with a (1.5 x 50) cm glass column.  The dimensions of the column were determined according to the following formulas :

diameter $= \sqrt[3]{m/10}$ cm    Where m is the amount of protein in mg

length = 30 x diameter

The gel was allowed to swell in an excess of 0.056 M Tris buffer, pH 7.(30 mL/gm gel) and left for 48 hours at 4°C.  Then the slurry was carefully poured into a vertical glass  column down the wall using a glass rod.  After the gel had settled, the column was equilibrated with Tris buffer pH 7.4 for 24 hours.

Void volume determination : The void volume was measured by blue dextran 2000 with a concentration of 1 mg/mL of Tris buffer (0.056 M, pH 7.4).  1 mL of blue dextran was applied to the column and eluted by the same buffer, with a flow rate of 20 mL/hour.  Fractions of  1 mL were collected and their absorbance was measured at 600 nm.

b.  Sample application and elution :

1 mL of crude CSF was applied to the surface of the gel and left for a few minutes until it is absorbed.  Elution was started by the addition of Tris buffer pH 7.4 and a number of fractions were collected, with a volume of 1 mL each at a flow rate of 20 mL/hour.  This was carried out in the cold room at 4°C.

c.  The activity and protein amount of each fraction was determined as described in section (2.2.2) and (2.2.1), respectively.

d.  The specific activity and fold of purification was determined.

4.  Ion exchange chromatography (pH gradient) :     Fractions collected by gel filtration, that showed activiy,were applied to a DEAE-Sephadex A-50 column for further separation of the isoenzymes.

**Materials :**

a.  Diethylaminoethyl (DEAE)-Sephadex A-50.

b.  Tris buffer (0.05 M) :   6.057 gm of Tris is dissolved in an appropriate amount of distilled water, then the volume is brought to 1M with distilled water.

c.  Tris-HCl (0.05 M) :   A number of Tris-HCl buffer solutions are prepared, each with a different pH.  This is done by mixing 50 mL of Tris buffer (0.05 M) with the amounts of 0.05M HCl illustrated below.  Then the volume is brought to 200 mL distilled water.

| pH, 25°C | (X) |
|----------|-----|
| 7.0 | 49.0 |
| 7.2 | 44.7 |
| 7.5 | 40.3 |
| 7.8 | 34.5 |
| 8.1 | 26.2 |
| 8.5 | 14.7 |
| 8.9 | 7.0 |

Where (X) refers to the amount of 0.05 M HCl (in mL) that was added to the 50mL of Tris buffer to obtain the corresponding pH.

## Procedure :

a. Gel preparation and column packing :

Ion exchange was carried out with a (2.2 x 21) cm glass column. 2.75 gm of the gel powder was allowed to swell in 250 mL of Tris-HCl buffer (0.05 M, pH 8.0) and left for 48 hours at 4°C. Then the slurry was carefully poured into a vertical glass column down the wall using a glass rod and filled to a height of 18 cm. After the gel had settled, the ion exchange column was equilibrated by pumping with Tris-HCl buffer pH 8.0.

b. Sample application and elution :

0.5 mL from each fraction collected by gel filtration, that showed activity, was taken and mixed together then concentrated about two times by the use of a dialysis bag against a solution of sucrose. Then 2 mL of the concentrated fractions was applied to the surface of the gel and left for a few minutes until it is absorbed. Elution was started as described in following steps. Fractions were collected with a volume of 2 mL each at a flow rate of 60 mL/hour. This was carried out in the cold room at 4°C :

1. 16 mL of Tris-HCl pH 8.9 was added and the fractions were collected in the first eight tubes.

2. 12 mL of Tris-HCl pH 8.5 was added and the fractions were collected in tubes number 9-14.

3. 10 mL of Tris-HCl pH 8.1 was added and the fractions were collected in tubes number 15-19.

4.     10 mL of Tris-HCl pH 7.8 was added and the fractions were collected in tubes number 20-24.

5.     14 mL of Tris-HCl pH 7.5 was added and the fractions were collected in tubes number 25-31.

6.     10 mL of Tris-HCl pH 7.2 was added and the fractions were collected in tubes number 32-36.

7.     20 mL of Tris-HCl pH 7.0 was added and the fractions were collected in tubes number 37-44.

c. The activity and protein of each fraction was determined as described in section (2.2.2) and (2.2.1), respectively.

d. The specific activity and fold of purification was determined.

### 2.2.3.2 Purification by ion exchange chromatography (salt gradient)

Another method was applied for the purification of LDH. The CSF sample was directly applied to an ion exchange column (DEAE-Sephadex A-50) and eluted by means of a salt gradient. For this purpose, different concentrations of sodium chloride were used. The process was carried out as follows :

**Materials :**

a.     DEAE-Sephadex A-50.

b.     Tris-HCl buffer (0.05 M, pH 8.0) : 50 mL of 0.05 M Tris (previously prepared by dissolving 6.055 gm of Tris in 1 L distilled water) is mixed with an appropriate amount of 0.05 M HCl to obtain a pH of 8.0. Then the volume is brought to 200mL with distilled water.

c.     Sodium chloride (NaCl) solution : Three different concentrations of NaCl solution were prepared (200, 150, 100) mM. A stock solution of 200 mM NaCl solution was prepared by dissolving 1.688 gm NaCl in 100 mL Tris-HCl buffer. The other two concentrations were prepared by serial dilution with the same buffer.

**Procedure :**

a.  Gel preparation and column packing :

   Ion exchange was carried out with a mini column (8 x 150 mm glass column). The column was plugged with a piece of glass wool placed at the bottom of the column. 90 mg of the gel powder was allowed to swell in 5.25 mL of Tris-HCl (0.05 M, pH 8.0) and left for 48 hours at 4°C. Then the slurry was carefully poured into the mini column using a glass rod and filled to a height of 4 cm. After the gel had settled, the column was equilibrated with 100 mM NaCl solution.

b.  The activity and protein of crude CSF was determined as described in section (2.2.2) and (2.2.1).

c.  Sample application and elution :

   0.25 mL of crude CSF was applied to the surface of the gel and left for a few minutes until it is absorbed. Elution was carried out stepwise with Tris-HCl buffer, containing successively (100, 150, 200) mM NaCl (pH 8.0 at 25°C). For each concentration, two 4 mL fractions were collected.

d.  The activity and protein of each fraction was determined as described in section (2.2.2) and (2.2.1).

e.  The specific activity and fold of purification was determined.

## 2.2.4 Separation of lactate dehydrogenase isoenzymes by polyacrylamide gel electrophoresis (PAGE)

   Electrophoresis was carried out using the LKB 2117 multiphor system and as follows :

**Materials :**

1.  Tris-glycine buffer stock solution pH 8.9 : 75.1 gm of glycine is dissolved in approximately 3 L distilled water. The solution is titrated with Tris to pH 8.9, then the volume is brought to 5 L with distilled water.

2.  Electrode buffer : 1 part of buffer stock solution is diluted with an equal amount of distilled water.

3.     Acrylamide solution :   22.2 gm acrylamide and 0.6 gm BIS are dissolved in 100 mL distilled water and stored in a dark bottle at 4°C.

4.     Ammonium persulphate :   150 mg ammonium persulphate is dissolved in 10mL distilled water and stored in a dark bottle at 4°C.

5.     Working solution :   This solution consists of the following :
       a. Tris buffer 0.056 M, pH 7.4.

       b. LiLactate 2 M : 4.8 gm LiLactate is dissolved in 25 mL distilled water.

       c. Phenazine methosulphate (PMS) : 1 mg of PMS is dissolved in 1 mL distilled water.  This kept at 4°C.

       d. Nitro blue tetrazolium (NBT) : 1 mg of NBT is dissolved in 1 mL distilled water.  This is kept at 4°C.

       e. NAD : 10 mg of NAD is added directly to the mixture.

       These constituents are mixed as follows :

       1 mL lactate
       3 mL NBT
       0.3 mL PMS
       1 mL Tris
       10 mL NAD

**Procedure :**

1.     Preparation of the gel with a concentration of 7.5% :   The gel solution is made by mixing the components in the proportions and order described below :

|                                      | mL   |
|--------------------------------------|------|
| a. Distilled water.                  | 7.5  |
| b. Tris-glycine buffer stock solution. | 33.0 |
| c. Acrylamide solution.              | 22.2 |
| d. Ammonium persulphate.             | 3.2  |
| e. TEMED.                            | 0.1  |

The solution is mixed carefully without introducing too much air. Then it is immediately poured into the molding set with the help of a plastic funnel with a pressed tip. The polymerization reaction will be completed in 40 minutes.

2.  The buffer tanks of the multiphor are filled with the electrode buffer, 1.2 L for each tank (the surface must not be higher than the top of the electrodes). The gel glass plate is placed on the cooling plate and connected with the buffer by means of 8-10 layers of electrode wicks. These are soaked in the buffer, then placed on the edge of the gel, covering 10-12 mm of it.

3.  The cooling water is switched on, the temperature should not exceed 10°C. The anticondensation lid is placed over and pre-electrophoresis is started. This is performed at 50 mA for 30 minutes.

4.  0.1 mL of the samples are applied and concentrated for 5-10 minutes with a current of 20 mA.

5.  The current is increased to 40 mA with a 15 v/cm field strength and left for nearly 90 minutes.

6.  After electrophoresis, the glass plate is removed and about 0.3 mL of working solution is put over the places where the samples were applied. Then the solution is spread homogeneously with a glass rod.

7.  The plate is then placed in the incubator for 30 minutes or more at 47°C, the purple bands should be seen.

8.  The plate is placed in the fixing solution for 10 minutes. This solution is prepared as follows :

    50% Methanol.
    40 mL distilled water.
    10% Glacial acetic acid.

## 2.2.5 Kinetic studies on LDH$_2$

Kinetic studies were carried out on LDH$_2$ isolated from CSF by ion exchange chromatography (salt gradient). These studies include the determination of initial rates, reaction order, the effect of different factors on the rate of the enzyme reaction and the inhibitory effect of both urea and oxalate. Activity measurements were carried out as described in described in section (2.2.2).

### 2.2.5.1 Effect of substrate concentration

**a. Sodium pyruvate :** From a stock solution of 20 mM the following concentrations were prepared by serial dilution with distilled water : (15, 13, 10, 7, 5, 2, 0.7, 0.5, 0.2) mM. The stock solution was prepared by dissolving 110 mg in 50 mL distilled water. The activity at each concentration was measured and the rate of reaction was plotted versus pyruvate concentration.

**b. Nicotineamide adenine dinucleotide (NADH) :** From a stock solution of 1 mM the following concentrations were prepared by serial dilution with Tris buffer (0.056 M, pH 7.5): (0.75, 0.25, 0.15, 0.10, 0.05) mM. The stock solution was prepared by dissolving 1.8 mg NADH in 25 mL Tris buffer (0.056 M, pH 7.5). The activity at each concentration was measured as described in section (2.2.2) using the optimum pyruvate concentration.

### 2.2.5.2 Effect of pH

Using optimum substrate concentration, the activity was measured at different pH (6.0, 6.5, 7.0, 7.5, 8.0). 1M HCl was used to adjust the pH of Tris buffer (0.056 M). The rate of the reaction was plotted versus the pH.

### 2.2.5.3 Effect of temperature

The effect of temperature on the rate of reaction was studied by measuring the activity at different temperatures (10, 15, 20, 25, 30, 35, 40, 45)°C, using optimum substrate concentration and pH. The rate of the reaction was plotted versus temperature.

### 2.2.5.4 Initial rate determination

The initial rate was determined as described in section (2.2.2.1). Four different concentrations of NADH were prepared (0.025, 0.05, 0.10, 0.25) mM. The change in each concentration was followed through out a period of time.

### 2.2.5.5 Effect of LDH$_2$ concentration on the initial rate

Three different concentrations of LDH$_2$ were taken (2.4, 3.6, 4.8) x10$^{-3}$ mg. The initial rate at each concentration was determined and plotted versus the enzyme concentration.

### 2.2.5.6 Reaction order

The reaction order was determined by applying the time-course data to many reaction rate equations, as will be described in the next chapter. The time course data were obtained by following the change in absorbance (concentration) through out 5 minutes at (10, 25, 30, 35)°C, using an NADH concentration of 0.75 mM and optimum conditions. Some of the rate equations used, covered the kinetics of the complex formation between LDH$_2$ and NADH, as well as the order of the reaction. These equations required the use of scatchard plots. The scatchard plots were obtained by following the reaction using different NADH concentrations (0.75, 0.25, 0.15, 0.10, 0.05) mM and at (10, 25, 30, 35)°C.

### 2.2.5.7 Inhibition of LDH$_2$

**a. Inhibition by urea :** The inhibitory effect of urea was studied by using different concentrations of urea (2, 1, 0.5, 0.25, 0.1) M at three different pyruvate concentrations (5, 2, 0.7) mM and optimum conditions. The different concentrations of urea were prepared by serial dilution with distilled water from a stock solution of 2 M. This was prepared by dissolving 3 gm urea in 25 mL distilled water. For the measurement of activity, the method described in section (2.2.2) was used except that 0.1 mL of urea was added after the addition of LDH$_2$ and incubated together.

**b. Inhibition by oxalate :** The inhibitory effect of oxalate was studied by using different concentrations of potassium oxalate (0.25, 0.125, 0.062, 0.031, 0.01) mM at three different pyruvate concentrations (5, 2, 0.7) mM and optimum conditions. The different potassium oxalate concentrations were prepared by serial dilution with distilled water from a stock solution of 0.25 mM. This was prepared by dissolving 0.1675 mg potassium oxalate in 50 mL distilled water. For the measurement of activity, the method described in section (2.2.2) was used except that 0.1 mL of oxalate was added after the addition of LDH$_2$ and incubated together.

### 2.2.6 Thermodynamic studies on LDH$_2$

The same experiments described in section (2.2.5.6) were carried out and the thermodynamic parameters of both the standard and transition state were obtained using Van't Hoff and Arrhenius equations, respectively. The details of these equations will be described in the next chapter.

### 2.2.7 Spectroscopic studies on LDH₂ and LDHₛ

Absorption measurements were made by an LKB ultrospec 4050 spectrophotometer in the U.V range (220-360) nm. Measurements were carried out using a 3 mL quartz cuvette with a 1 cm path length. The molar absorption coefficient was determined from Lambert-Beers law, which relates the absorbance with the concentration according to the following equation :

$$A = \varepsilon l C$$   Where $\varepsilon$ is the molar absorption coefficient

### 2.2.7.1 The U.V spectrum of LDH₂ and LDHₛ

**a.**   The U.V spectrum of LDH₂ :  2.5 mL of the fraction containing LDH₂ (pH 8.0), was placed in the cuvette and the absorbance was measured in the range of (220-360)nm at 30°C.

**b.**   The U.V spectrum of LDHₛ :  2.5 mL of the fraction containing LDHₛ (pH 8.0) was placed in the cuvette and the absorbance was measured in the range of (220-360)nm and at 30°C.

### 2.2.7.2 Factors affecting the absorption properties of LDH₂ and LDHₛ

**a.**   **Temperature effect on the U.V spectrum of LDH₂ and LDHₛ :**

1.   Temperature effect on the U.V spectrum of LDH₂ :  2.5 mL of the fraction containing LDH₂ (pH 8.0) was placed in the cuvette and incubated for 10 minutes at 45°C.  The absorbance was measured in the range of (220-360)nm.

2.   Temperature effect on the U.V spectrum of LDHₛ :  2.5 mL of the fraction containing LDHₛ (PH 8.0) was placed in the cuvette and incubated for 10 minutes at 45°C.  The absorbance was measured in the range of (220-360) nm.

**b.**   **PH effect on the U.V spectrum of LDH₂ and LDHₛ :**

1.   PH effect on the U.V spectrum of LDH₂ :  The effect of changing the pH to 7.0 was studied by taking 2.5 mL of the fraction containing LDH₂ (pH 7.0) and measuring the absorbance in the range of (220-360) nm at 30°C.

2.   PH effect on the U.V spectrum of LDHs :   The effect of changing the pH to 7.0 was studied by taking 2.5 mL of the fraction containing LDHs (pH 7.0) and measuring the absorbance in the range of (220-360) nm.

### 2.2.7.3  Binding studies on LDHs with NADH

The spectral changes that accompany the binding of an enzyme to certain compounds was studied on LDHs with NADH.  A spectrum of the LDHs -NADH complex was obtained and many factors that affect this spectrum were studied.

### a.  The absorption spectrum of the LDHs-NADH complex

2.250 mL of NADH (0.150 mM, pH 8.0) was placed in the cuvette, then 0.250mL of the fraction containing LDHs was added.  The mixture was incubated for 30 minutes at 30°C and the absorbance was measured in the range of (220-360nm).

### b.  PH effect on the spectrum of the LDHs-NADH complex

0.150 mM NADH was prepared at five different pH (6.0, 6.5, 7.0, 7.5, 8.0). Then 2.250 mL of NADH, for each pH, was placed in the cuvette and 0.250 mL of the fraction containing LDHs was added.  The mixture was incubated for 30 minutes at 30°C and the absorbance was measured in the range of (220-360).

### c.  LDH5 : NADH concentration effect on the spectrum of the LDH5-NADH complex

Different amounts of both LDHs and NADH were mixed together. (0.250,2,0.750,1)mL of the fraction containing LDHs (pH 6.5) was added to (2.250, 2, 1.750, 1.50) mL of 0.150mM NADH (pH 6.5), respectively.  The mixture was incubated, for each concentration, for 30 minutes at 30°C and the absorbance was measured in the of (220-360) nm.

### d.  The stability of the LDHs-NADH complex

The stability of the LDHs -NADH complex was followed by measuring the absorbance of the complex through out 50 days.  The complex was stored at 4°C and the absorbance was measured in the range of (220-360) nm at 30°C.

# CHAPTER THREE

## RESULTS
## &
## DISCUSSION

## 3.1 Lactate dehydrogenase studies

In this study, measurements of different CSF constituents were carried out on 43 CSF specimen. CSF was obtained by lumbar puncture from patients with different neurological diseases. The host information of the 43 patient are illustrated in table (3.1)

Table (3.1).    The host information of patients with different neurological diseases.

| Disease | No. of patients | Age / (MAD*) |
|---|---|---|
| Bacterial meningitis | 15 | 9 ± 7  years |
| Hydrocephaly | 14 | 5.8 ± 4.7  months |
| Trauma | 5 | 13.6 ± 6.6  years |
| Meningocele | 3 | 18.7 ± 14.23  years |
| subdural effusion | 2 | 10 ± 2  months |
| Benign intracranial hypertension | 2 | 37.5 ± 5  years |
| Brain tumor | 2 | 20 ± 2  years |
| Total | 43 | |

* Mean average deviation.

## 3.1.1 Cerebrospinal fluid total protein measurement

The normal value of lumbar CSF protein is in the range of 15-45 mg/dL [27]. This value has been found to increase in some pathological conditions of the central nervous system.  Table (3.2) shows the protein concentration of the 43 patient suffering from different neurological diseases.  Increased concentration of lumbar CSF protein, is attributed to many factors.  Generally, the diseases that affect the permeability of the blood-CSF barrier cause this elevation.  Such diseases increase the permeability of the this barrier and so the ultrafiltration of plasma across this barrier will increase [1], [90].  However, in subdural effusion the cause of high protein concentration is attributed to leakage of plasma proteins [99].

Table (3.2).    Total protein concentration in CSF from patients with different neurological diseases.  The Lowry method was used for the determination of total protein concentration.  Details are described in section (2.2.1).

| Disease | No. Of patients | Protein concentration (mg/dL) | |
|---|---|---|---|
| | | Mean | Range |
| Bacterial meningitis | 15 | 627 | 100 - 3150 |
| Hydrocephaly | 14 | 40 | 12 - 50 |
| Trauma | 5 | 68 | 56 - 81 |
| Meningocele | 3 | 460 | 70 - 900 |
| subdural effusion | 2 | 528 | |
| Benign intracranial hypertension | 2 | 41 | |
| Brain tumor | 2 | 680 | |
| Standard | | | 15 - 45 |

**Immunological measurements** : The quantitative partition of CSF proteins by immunological methods demonstrates the presence of most serum proteins, such as prealbumin, albumin and immunoglobulins; IgG, IgA and IgM.  The source of proteins in CSF is either from diffusion across the blood-CSF barrier or synthesis within the central nervous system.  Albumin is not produced within the central nervous system and so immunological measurements of CSF albumin reflects the integrity of the blood-CSF barrier (99).  The plasma/CSF ratio gives very useful information about this barrier, an increased ratio may be due to an increased permeability of the blood-CSF barrier, or to impaired resorption of CSF protein caused by meningitis.  The normal plasma/CSF ratio is found to be 23.6 mg/dL (15).  However, IgG is also important for such purpose and for evaluating the synthesis of IgG within the central nervous system.  IgA and IgM are also important but to a lesser extent (99).  Table (3.3) shows the albumin, IgG and IgA amounts in eight patients suffering from different pathological conditions of the central nervous system.

Table (3.3).   Concentration of proteins in CSF of eight patients suffering from different neurological diseases.

| Patient No. | Disease | Total protein (mg/dL) | Albumin (mg/dL) | IgG (mg/dL) | IgA (mg/dL) |
|---|---|---|---|---|---|
| 1 | Bacterial meningitis | 206 | 26 | 6 | 3 |
| 2 | Trauma | 56 | 28 | 7 | 1 |
| 3 | Brain tumor | 780 | 27 | 11 | 2 |
| 4 | Meningocele | 900 | 28 | 6 | 4 |
| 5 | Benign intracranial hypertension | 35 | 18 | 8 | 2 |
| 6 | subdural effusion | 950 | 24 | 9 | 2 |
| Standard* | | 15 - 45 | 15.5 | 1.23 | 0.13 |

*Source: reference (15), table (19.3), p.468 .

### 3.1.2  Activity measurements of lactate dehydrogenase in cerebrospinal fluid

CSF lactate dehydrogenase has appeared to be of diagnostic value in many neurological diseases, especially in bacterial meningitis (34). The normal value of CSF LDH activity is in the range of 7-30 U/L (78). This level has been found to increase in some pathological conditions of the central nervous system (34). The values of LDH activity of the 43 patient suffering from different neurological diseases are illustrated in table (3.4). An increase in activity was found in meningitis, meningocele, subdural effusion and brain tumor.

The rise in LDH activity in the patients in the cases of bacterial meningitis, comes in agreement with other studies. Most of the rise in activity is due to the isoenzymes LDH$_4$ and LDH$_5$ which are derived from granulocytes (90), (91), (100). However, it is found that if encephalitis is accompanied with meningitis, the rise in LDH activity is due to the isoenzymes LDH$_1$ and LDH$_2$. This is attributed to the brain tissue itself (100). In subdural effusion, the increase in activity is attributed to all five isoenzymes. The findings suggest that in cases of hemorrhage, blood reaches CSF and causes this increase in LDH activity, especially that the LDH activity in plasma is nearly ten times greater than that in CSF (15),(34). The rise in activity found in the patients harboring a benign brain tumor, comes in agreement with many studies. In benign primary CNS tumors, an inconsistent elevation LDH activity is found.

While malignant primary and metastatic tumors show a more uniform rise in LDH activity. However, the cause of elevation is due to the secretion of LDH by the tumor cells (52). In malignant tumors the rise in activity is due to $LDH_4$ and $LDH_5$, because during the course of malignant transformation cells derive a great amount of energy from the anaerobic process. As a result, the activity of the anaerobic isoenzymes $LDH_4$ and $LDH_5$ increases (93). Another reason for the elevation of LDH activity in the previous cases, could be the permeability of the blood-CSF barrier. Such diseases affect the permeability of this barrier, which leads to an increase in the ultrafiltration of plasma across the barrier (90). To this extent, the normal value of LDH activity found in trauma, benign intracranial hypertension and hydrocephaly means that this barrier has not been affected.

Table (3.4). Total LDH activity in CSF from patients with different neurological diseases. The Scandinavian method was used to assay total CSF LDH. Details are described in section (2.2.2).

| Disease | No. Of patients | LDH activity (U/L) | |
|---|---|---|---|
| | | Mean | Range |
| Bacterial meningitis | 15 | 183 | 70 - 468 |
| Hydrocephaly | 14 | 31 | 6 - 104 |
| Trauma | 5 | 31 | 12 - 56 |
| meningocele | 3 | 106 | 77 - 150 |
| subdural effusion | 2 | 173 | |
| Benign intracranial hypertension | 2 | 19 | |
| Brain tumor | 2 | 85 | |
| Standard* | | | 7 - 30 |

*The value is for the pyruvate to lactate direction and at 30°C (78).

### 3.1.3 Purification of cerebrospinal fluid lactate dehydrogenase isoenzymes

### 3.1.3.1 purification by gel filtration and ion exchange (pH gradient)

Relevent studies on LDH in CSF do not refer to any purification process of this enzyme in CSF. Such studies are limited to activity measurements in different neurological diseases. In this study, the purification process of CSF LDH isoenzymes included two steps. The first step was the separation of CSF LDH by gel filtration on a Sephadex G-150 (1.5 x 50) cm column. The second step was applying the separated enzyme to a DEAE-Sephadex A-50 (2.2 x 21) cm anion exchange column and as follows :

a. **Gel filtration** : The CSF specimen applied to the gel filtration column, was from a patient with bacterial meningitis. The void volume (Vo) was determined and found to be equal to 31 mL at a flow rate of 20 mL/hour. As shown in figure (3.1), LDH activity eluted as two peaks. The first peak (peak 1) represents fractions number 17-21, while the second peak (peak 2) represents fractions number 28-36. 0.5 mL of each fraction representing peak 1 was taken and mixed together, then concentrated about two times by the use of a dialysis bag against a solution of sucrose. The same was done for peak 2. The activity of both concentrated fractions representing peak 1 and peak 2 was measured. Table (3.5) summarizes the results of gel filtration. The table shows that the total LDH activity of peak 1 and peak 2 is nearly equal to the LDH activity of crude CSF. This means that all LDH was eluted through the column. This is also true for the total yield of the two peaks which is equal to 98%, which means that the total enzyme activity was retained through the separation by gel filtration. The protein concentration of peak 1 was found to be 50 mg/L, while that of peak 2 was $1.3 \times 10^3$ mg/L. The specific activity specific activity of both peaks was also calculated an was found to be 0.3 U/mg protein for peak 1 and 0.041 U/mg protein for peak 2. A difference in the purification factor was found between the two peaks. The first peak was purified 130 time while the second peak was purified 18 time only. This could be attributed to the high protein concentration of the second peak.

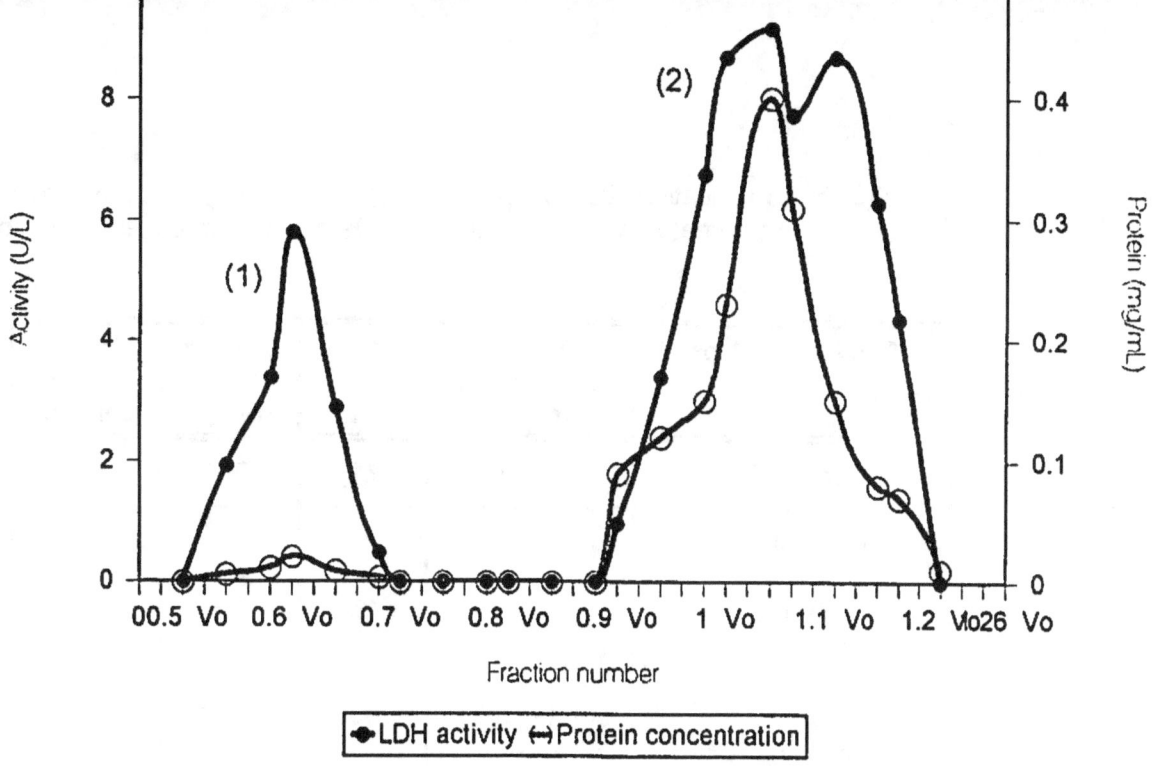

Figure (3.1). The elution profile of LDH from a Sephadex G-150 column (1.5 x 50 cm). Details are described in section (2.2.3.1).

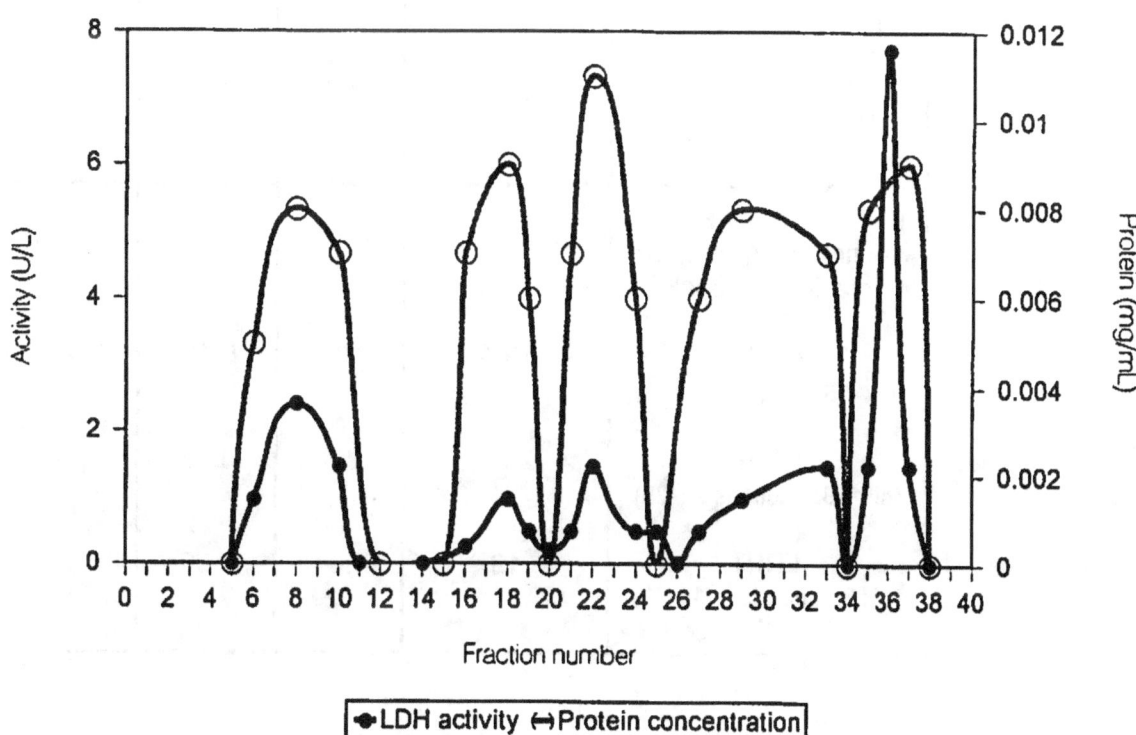

Figure (3.2). The elution profile of LDH isoenzymes from a DEAE-Sephadex column (2.2 x 21 cm). Details are described in section (2.2.3.1).

Table (3.5).    Purification of LDH isoenzymes from CSF by gel filtration and ion exchange chromatography (pH gradient). Details are described in section (2.2.3.1) .

| Purification steps | Protein concentration (mg/L) | Activity (U/L) | Specific activity (U/mg) | Yield (percent) | Fold of purification |
|---|---|---|---|---|---|
| 1. Crude CSF | $31.5 \times 10^3$ | 71 | 0.0023 | 100 | 1 |
| 2. Gel filtration : | | | | | |
| **Peak 1** (fractions 17-21) | 50 | 15 | 0.3 | 21 | 130 |
| **Peak 2** (fractions 28-36) | $1.3 \times 10^3$ | 55 | 0.041 | 77 | 18 |
| 3. Ion exchange chromatography(pH gradient) of peak 1: | | | | | |
| **LDH$_1$** (pH 7.0, fractions 35-38) | 16.7 | 9.7 | 0.6 | 14 | 261 |
| **LDH$_5$** (pH 8.9, fractions 5-12) | 18.3 | 4.8 | 0.26 | 7 | 113 |
| 4. Ion exchange chromatography(pH gradient) of peak 2 : | | | | | |
| **LDH$_2$** (pH 7.5, fractions 27-33) | 19.4 | 3 | 0.15 | 4 | 65 |
| **LDH$_3$** (pH 7.8, fractions 21-25) | 20.8 | 2.4 | 0.12 | 3 | 52 |
| **LDH$_4$** (pH 8.1, fractions 15-20) | 19 | 1.44 | 0.08 | 2 | 35 |

**b. Ion exchange (pH gradient) :** Anion exchange chromatography is one of the methods used for the separation of LDH isoenzymes. The principle of this method is the different total charge each isoenzyme has. Therefore, lowering the pH of the anion exchanger will gradually weaken the binding of the adsorbed isoenzymes. This will result in the separation of each isoenzyme at a certain pH (101).

The anion exchange method used in this study is the same as that in a previous study by Akrawi B.A. (102). 2 mL of the concentrated fractions representing peak 1 (separated by gel filtration) was applied to the DEAE-Sephadex A-50 column. Only the fractions at pH 8.9 and pH 7.0 showed activity. In between this range, no activity was found. From the other concentrated fractions representing peak 2, 2 mL was taken and applied to the anion exchange column. The fractions that showed activity were at pH (8.1, 7.8, 7.5). As a result, five peaks were obtained; at pH (8.9, 8.1, 7.8, 7.5, 7.0). These peaks are illustrated in figure (3.2). These peaks represent $LDH_5$, $LDH_4$, $LDH_3$, $LDH_2$ and $LDH_1$, respectively.

The elution of $LDH_5$ at the beginning of the process, comes from the fact that at pH 8.9 $LDH_5$ has a negative charge and so it elutes freely without being adsorbed by the anion exchanger. Whereas, $LDH_1$ and at pH8.9 has the highest negative charge, so it is strongly adsorbed by the anion exchanger and will not elute until the pH reaches 7.0. However, electrophoretic identification of column fractionated LDH was not possible, due to the very low activity of the isoenzymes. The results of ion exchange chromatography are summarized in table (3.5). The values of activity of the isoenzymes show that most of the total LDH activity was lost through out the anion exchange process, though a good purification factor was obtained for each isoenzyme. The protein concentration for the five isoenzymes, from $LDH_1$ to $LDH_5$ were (16.7, 19.14, 20.8, 19, 18.3) mg/L, respectively and the specific activity was found to be (0.6, 0.15, 0.12, 0.08, 0.26) U/mg protein, respectively.

Because of the low activity of the isoenzymes obtained by this method, another anion exchange method was carried out by the use of a salt gradient, as described in the following section.

### 3.1.3.2  Purification by ion exchange chromatography (salt gradient)

The method of Donald Mercer was used for the purification of CSF LDH isoenzymes (103). Crude CSF was applied to a DEAE-Sephadex A-50 mini column (18 x 150) mm. Elution was carried out using 50 mM Tris (hydroxymethyl) methyl amine hydrochloride, containing successively (100, 150, 200) mM sodium chloride. Two 4 mL fractions of each concentration was collected.

$LDH_5$ eluted at 100 mM NaCl and in the first 4 mL fraction. The second 4 mL fraction showed no activity. LDH$_1$, LDH$_2$ and LDH$_3$ eluted in the first 4 mL fraction at 150 mM NaCl. In the second 4 mL fraction, no activity was found. At 200 mM NaCl and in the first and second 4 mL fractions, LDH$_2$ was eluted. In the first 4 mL (200 mM NaCl), a carryover of LDH3 was found. However, in the second 4 mL fraction only LDH$_2$ was eluted. According to Mercer's method, LDH$_5$ eluted with LDH$_4$ and LDH$_3$ at 100 mM NaCl. LDH$_1$ and LDH$_1$ eluted at (150, 200) mM, respectively.

Results of the purification of LDH$_2$ and LDH$_5$ are summarized in table (3.6). The table shows that the activity of the purified LDH$_2$ was 13 U/L with a protein concentration of 24 mg/L, while that for LDH$_5$ was 10 U/L with a protein concentration of 79.1 mg/L. The yield of LDH$_2$ was 6%, while that for LDH5 was 5%. LDH$_2$ was purified 16.4 times, while LDH$_5$ was purified 4 times.

Table (3.6).  Purification of LDH$_2$ and LDH$_5$ from CSF by anion exchange chromatography by means of a salt gradient. Details are described in section (2.2.3.2).

| Fraction | Protein concentration (mg/L) | Activity (U/L) | Specific activity (U/mg) | Yield (Percent) | Fold of purification |
|---|---|---|---|---|---|
| Crude | 6.6 x 10$^3$ | 217 | 0.033 | 100 | 1 |
| LDH$_2$ (200 mM NaCl) | 24 | 13 | 0.54 | 6 | 16.4 |
| LDH$_5$ (100 mM NaCl) | 79.1 | 10 | 0.13 | 5 | 4 |

By comparing these results with those obtained from anion exchange via pH gradient, a better yield and purification factor was found in the anion exchange via pH gradient. Despite this fact, the activity of LDH$_2$ and LDH$_5$ obtained via salt gradient was higher than that obtained by pH gradient. For this reason, LDH$_2$ and LDH$_5$ isolated from anion exchange chromatography by means of a salt gradient were used for the kinetic, thermodynamic and spectroscopic studies.

### 3.1.4 Electrophoretic pattern of lactate dehydrogenase isoenzymes in the cerebrospinal fluid

CSF LDH isoenzymes were characterized by PAGE, the bands were detected by the use of a substrate, containing lactate, NAD, phenazine methosulphate (PMS) and nitro blue tetrazolium(NBT). The purple bands representing the isoenzymes, were developed according to the following reaction (47) :

$$\text{Lactate} + \text{NAD}^+ \xrightleftharpoons{\text{LDH}} \text{Pyruvate} + \text{NADH} + \text{H}^+$$

NADH then reduces the tetrazolium salt to a purple formazine compound (NBF) :

$$\text{NADH} + \text{NBT} + \text{PMS} \longrightarrow \text{NAD}^+ + \text{NBF}$$

Figures (3.3) and (3.4) show the electrophoretic pattern of CSF LDH isoenzymes in some of the neurological diseases. Figure (3.5 a, b) shows the electrophoretic pattern of the isoenzymes $LDH_2$ and $LDH_5$ isolated from CSF from a patient with bacterial meningitis, by anion exchange chromatography via salt gradient.

CSF LDH was separated into five isoenzymes (figures 3.3, 3.4) with different amounts, depending on the type of disease. In subdural effusion the predominant isoenzymes were $LDH_1$, $LDH_2$ and $LDH_3$. Findings suggest that in subdural effusion, the activity of all isoenzymes increase. Because in such cases blood reaches CSF. In meningocele, the predominant isoenzyme was found to be $LDH_2$. The literature does not refer to any electrophoretic pattern in cases of meningocele, but usually the predominance of the fast moving isoenzymes is attributed to the brain tissue (100). In the bacterial meningitis cases, different patterns were obtained. In figure (3.3), the bacterial meningitis cases showed a predominance of $LDH_1$, $LDH_2$ and $LDH_3$. Whereas, figure in (3.4) the bacterial meningitis cases showed a predominance of $LDH_5$. This is attributed to the degree of infection. If the infection reaches the brain, causing encephalitis, the predominant isoenzymes will be LDH1 and LDH2. When the brain is not involved, the predominant isoenzyme is LDH5 which is derived from the granulocytes (100). In four cases of hydrocephaly, a uniform pattern was found (figure 3.4). The predominant isoenzymes, and the only ones seen in the pattern, were $LDH_1$, $LDH_2$ and $LDH_3$. As previously mentioned, fast moving isoenzymes are attributed to the brain tissue.

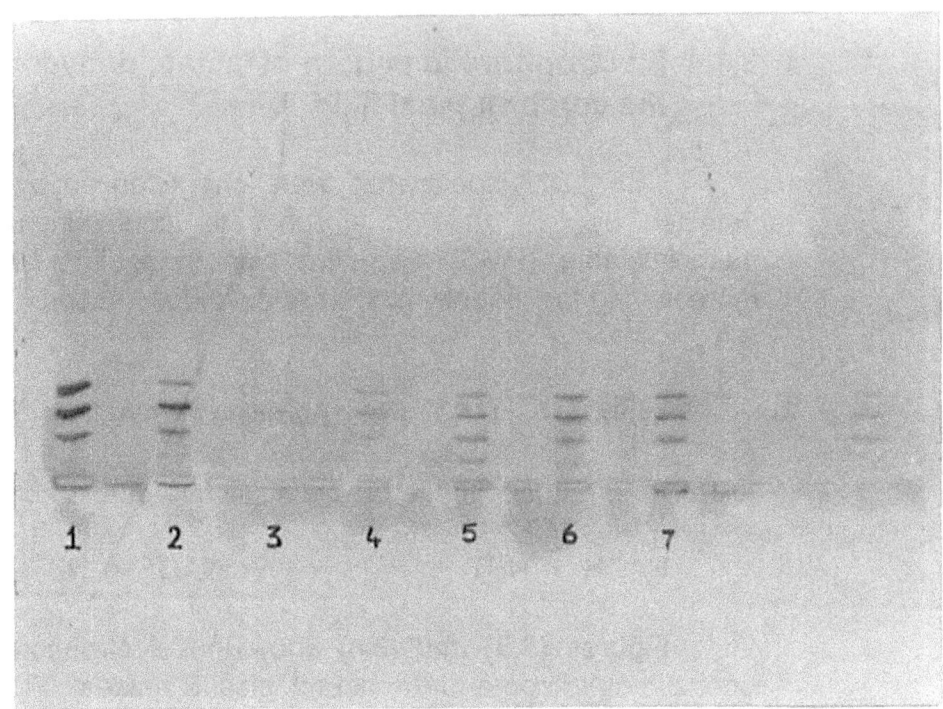

Figure (3.3).   Conventional PAGE (7.5%) pattern of CSF LDH isoenzymes in cases of subdural effusion (1), meningocele (2) and meningitis (5,6,7). Details are described in section (2.2.4).

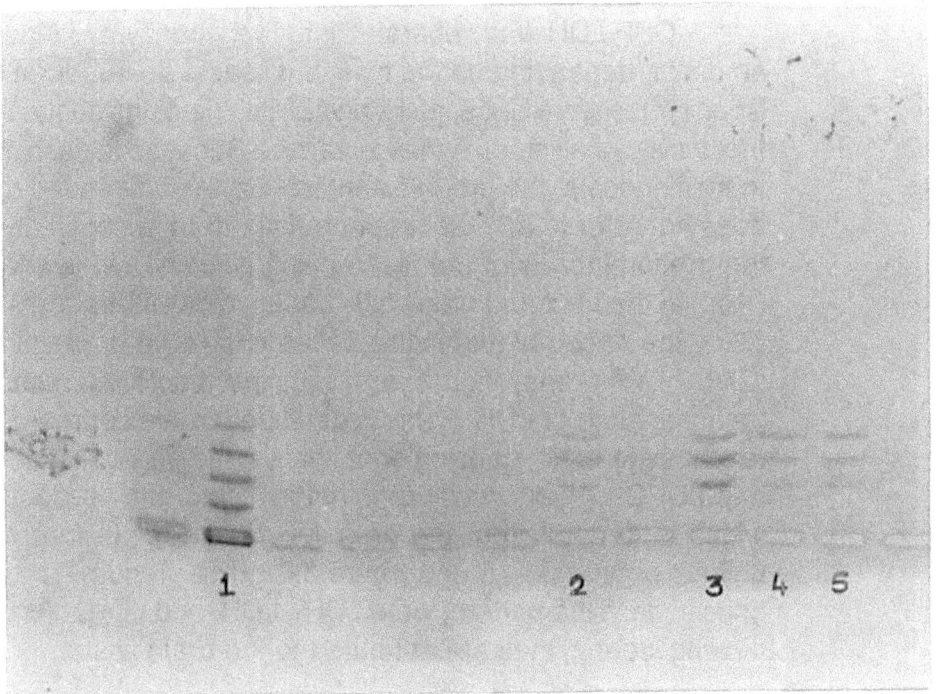

Figure (3.4).   Conventional PAGE (7.5%) pattern of CSF LDH isoenzymes in cases of bacterial meningitis (1) and hydrocephaly (2,3,4,5). Details are described in section (2.2.4).

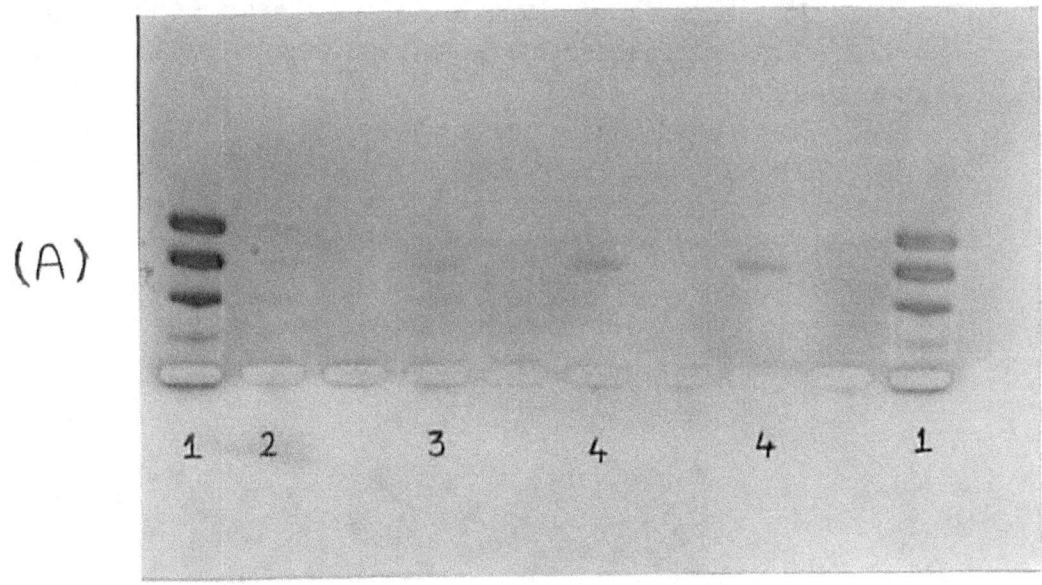

(A)

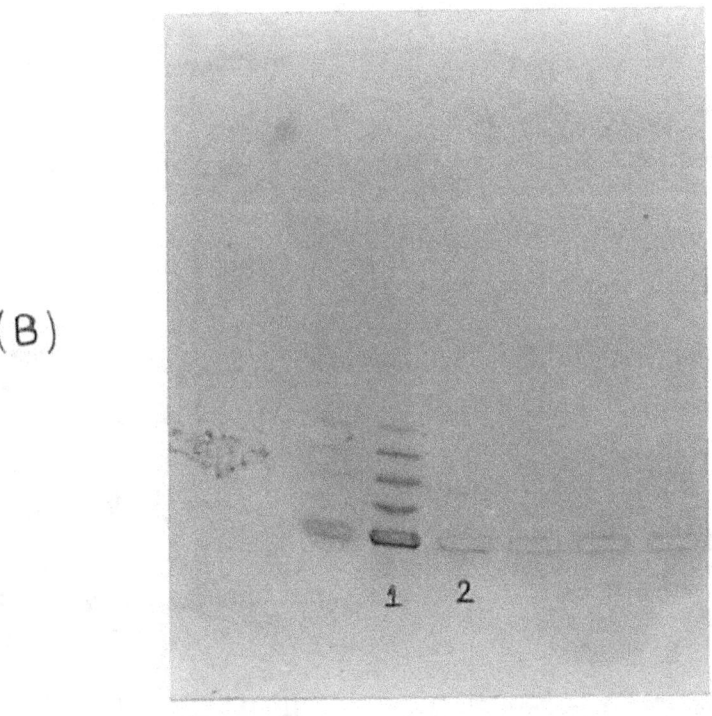

(B)

Figure (3.5).Conventional PAGE (7.5%) pattern of CSF LDH isoenzymes separated by anion exchange chromatography via salt gradient. **(A)**: column (1) represents the crude CSF sample. (2) represents $LDH_1$,$LDH_2$ and $LDH_3$ eluted at 150 mM NaCl. (3),(4) represent $LDH_2$ eluted at 200 mM NaCl. **(B)**: column (2) represents $LDH_5$ eluted at 100 mM NaCl. Details are described in section (2.2.4).

In a bacterial meningitis case, the electrophoretic pattern of CSF LDH revealed an extra isoenzyme which was designated LDHs (figure 3.6). LDHs migrated faster than LDH1, which means that its charge is more negative than $LDH_1$. Consequently, LDHs has a different structure from $LDH_1$. Such extra bands have been reported by many investigators, in electrophoretic patterns of serum LDH in many cases of human cancer (55),(56), (59), (61). These extra bands or extra isoenzymes had different electrophoretic mobilities. One of these extra isoenzymes, is that reported by Lubin, J. et al. designated $LDH_1$ ex (56). It migrated faster than $LDH_1$. The extra isoenzyme found in the meningitis case (LDHs), resembles $LDH_1$ ex in its electrophoretic mobility.

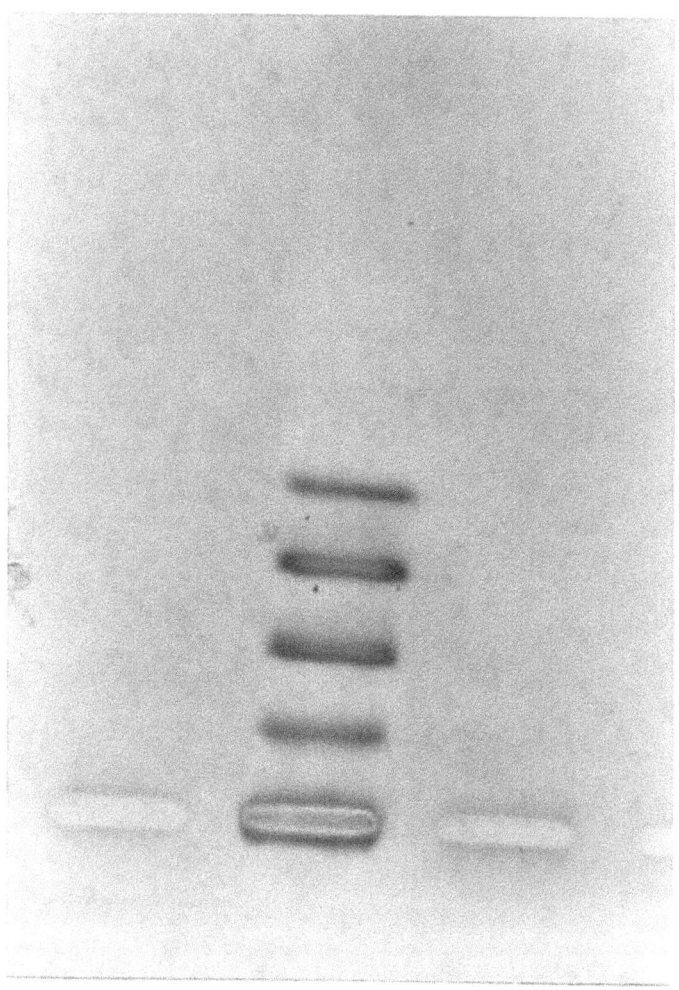

Figure (3.6). Conventional PAGE (7.5%) pattern of CSF LDH isoenzymes from a patient with bacterial meningitis, showing the extra band (LDHs). Details are described in section (2.2.4).

### 3.1.5 Kinetic studies on $LDH_2$

### 3.1.5.1 Effect of substrate concentration

**Sodium pyruvate and NADH :** Figure (3.7) and (3.8) show the effect of pyruvate and NADH on the velocity of the reaction, respectively. The reults indicated that the optimum pyruvate concentration was 5 mM, while that for NADH was 0.75 mM. The curve obtained by plotting the velocity described as U/L versus pyruvate or NADH concentration gave a right rectangular hyperbola. Different ratios of pyruvate and NADH concentrations for two fractions of $V_{max}$, were obtained in order to test the obedience of $LDH_2$ for the Michaelis-Menton kinetics. Table (3.7) shows the theoretical and obtained values of different ratios of pyruvate and NADH concentrations. The values of these ratios appeared to be close to the theoretical ones. Hence, $LDH_2$ obeys Michaelis-Menton kinetics.

Table (3.7).  Theoretical and obtained values for different ratios of substrate concentration at two fractions of $V_{max}$.

| [S] / [S] ratio | Theoretical | Obtained | |
|:---:|:---:|:---:|:---:|
| | | Pyruvate | NADH |
| $[S]_{0.9} / [S]_{0.5}$ | 9 | 7.6 | 5.7 |
| $[S]_{0.9} / [S]_{0.1}$ | 81 | 53 | 103 |
| $[S]_{0.75} / [S]_{0.5}$ | 3 | 2 | 2.3 |
| $[S]_{0.8} / [S]_{0.5}$ | 4 | 3 | 3 |

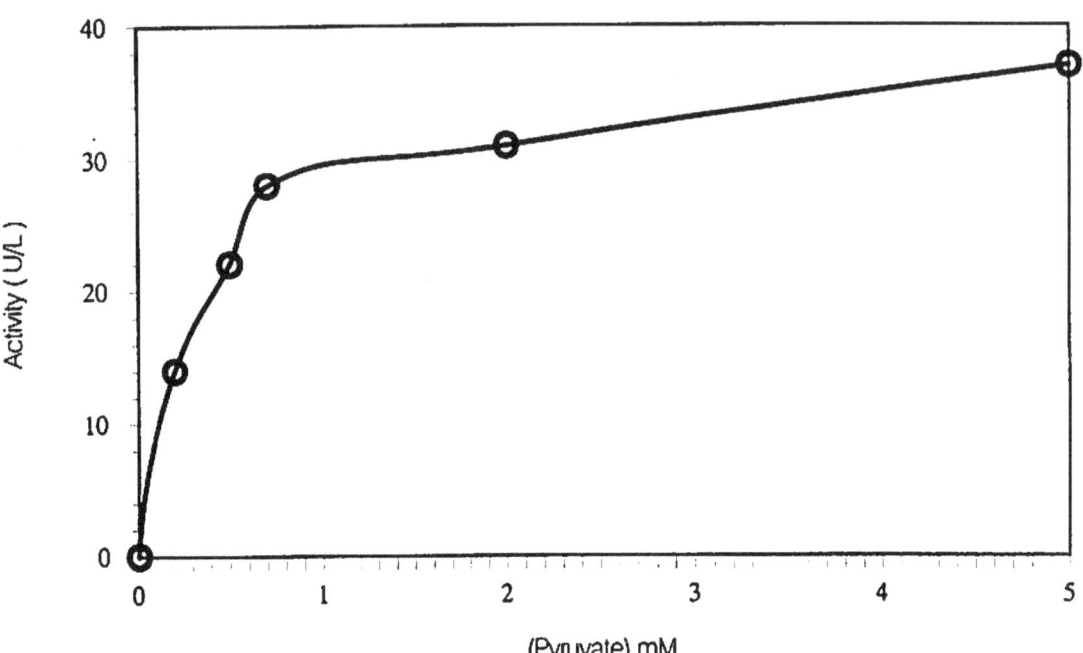

Figure (3.7). The effect of pyruvate concentratin on the rate of reaction. Details are described in section (2.2.5.1).

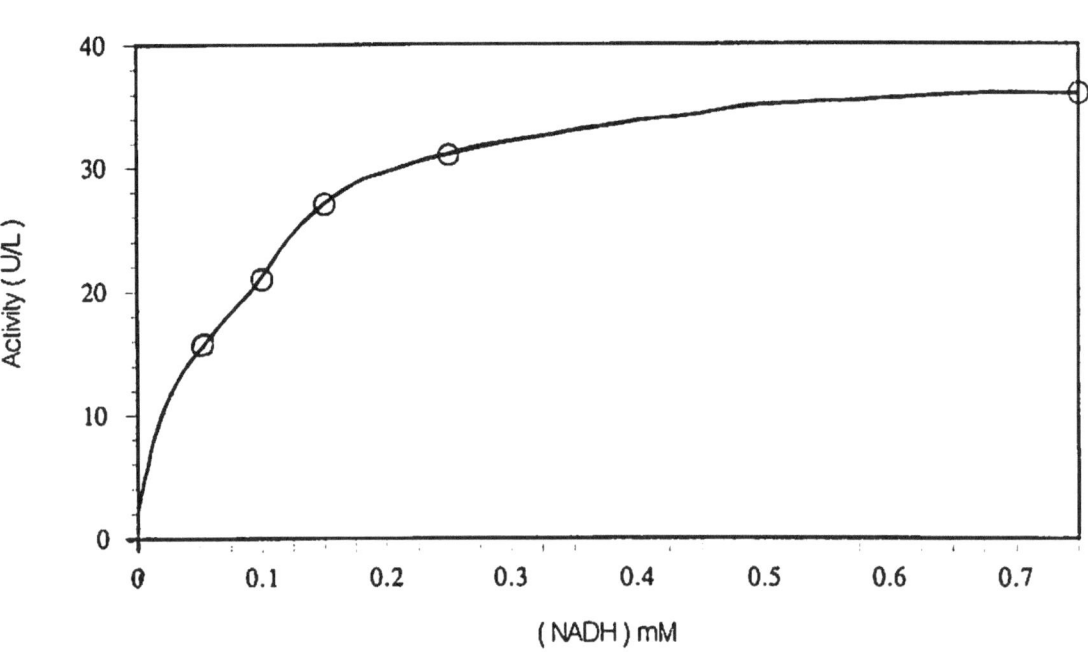

Figure (3.8). The effect of NADH concentration on the rate of reaction. Details are described in section (2.2.5.1).

### 3.1.5.2  Effect of pH

It is known that the pH of an enzyme catalyzed reaction, has a great effect on the velocity of the reaction.  This arises from the fact that the active sites in enzymes, are composed of ionizable groups that have a certain ionic form, where the catalysis of the reaction is maintaned.  In addition, the substrate itself may have ionizable groups and only at a certain pH the substrate will have an ionic form where it can bind to the enzyme (104), (105).  The eefect of pH on the velocity of the reaction is shown in figure (3.9), which represents a plot between the vlocity of the reaction described as U/L and the pH.  As illustrated in the figure, the velocity of the reaction increases with increasing the pH until it reaches 7.0 (pH optimum).  Then the velocity declines above this pH.

### 3.1.5.3  Effect of temperature

The catalytic activity of an enzyme results from the ordered tertiary structure of the enzyme.  This enfluences the stereospecificity of the binding sites.  A number of noncovalent bonds participate in maintaining the tertiary structure of the enzyme.  Which means that the this structure is easily affected if high temperatures are used.  Figure (3.10) shows the effect of temperature on the velocity of the reaction.  As shown in the figure, raising the temperature above the optimum degree, which was 35°C, causes a decline in the velocity, indicating a disruption of the tertiary structure.

### 3.1.5.4  Effect of enzyme concentration on the initial velocity

The initial velocity at any substrate concentration is given by :

$$v = \frac{[S] \, K_p}{K_m + [S]} \, [E]_t$$

This equation illustrates the direct proportion of the initial velocity to the total enzyme concentration.  Figure (3.11) reflects this fact.  Different concentrations of $LDH_2$ were plotted versus the initial velocity described as μM/min, and a direct proportin was obtained.

### 3.1.5.5  Determination of $K_m$ for NADH

The $K_m$ value is an important constant in enzyme sudies.  It represents the substrate  concentration at which the initial velocity is half maximal (106).  The Lineweaver-Burk plot was used for the determination of $K_m$ value for NADH :

$$\frac{1}{v} = \frac{K_m}{V_{max}} \frac{1}{[S]} + \frac{1}{V_{max}}$$

Figure (3.12) represents the Lineweaver-Burk plot. The reciprocal of initial velocity described as µM/min, was plotted versus the reciprocal NADH concentration in µM. A straight line was obtained, with a slope of $K_m/V_{max}$, an intercept of $1/V_{max}$ on the $1/v$ axis and an intercept of $(-1/k_m)$ the $1/[S]$ axis. The value of $k_m$ was found to be 100 µM, while $V_{max}$ was found to be 2.2 µM/min.

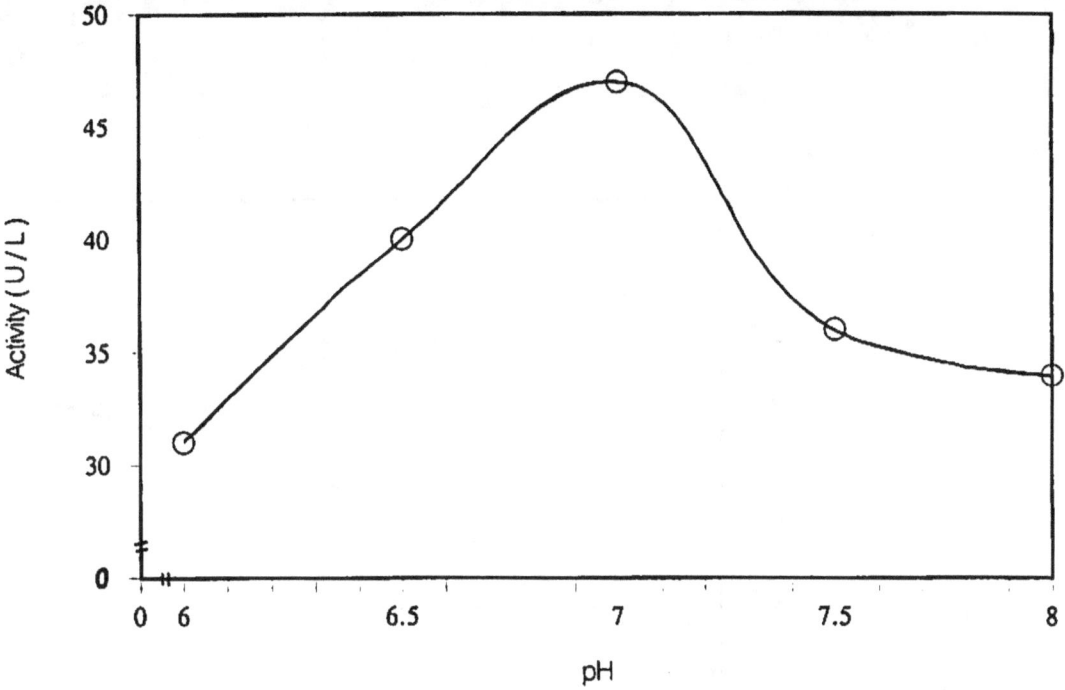

Figure (3.9). The effect of pH on the rate of reaction. Details are
described in section (2.2.5.2).

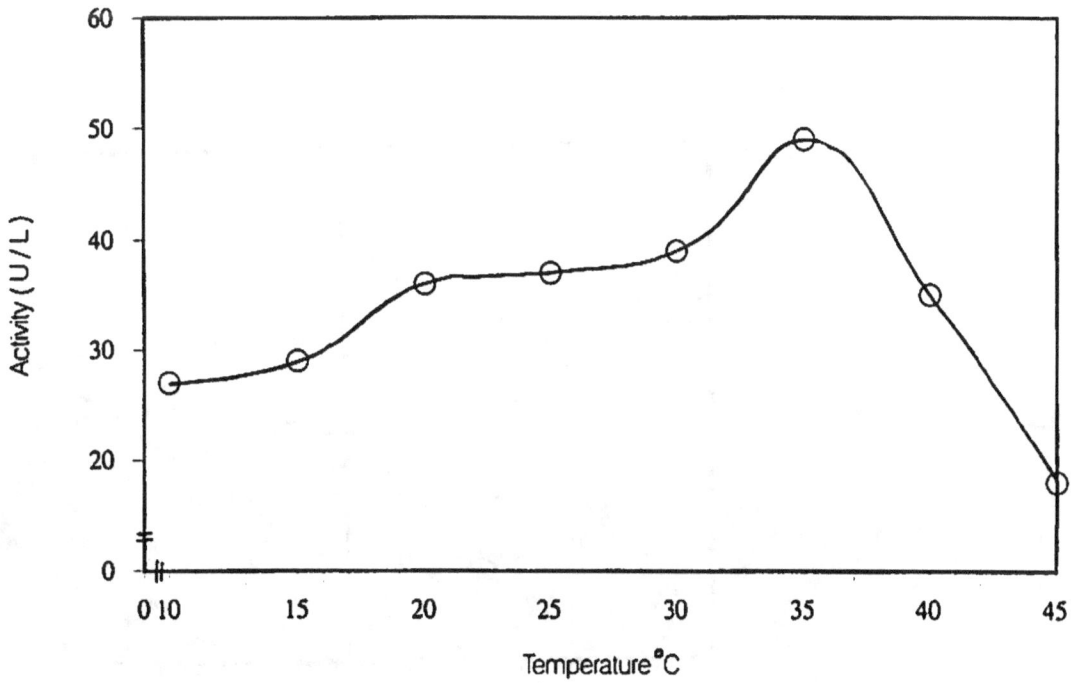

Figure (3.10). The effect of temperature on the rate of reaction.
Details are described in section (2.2.5.3).

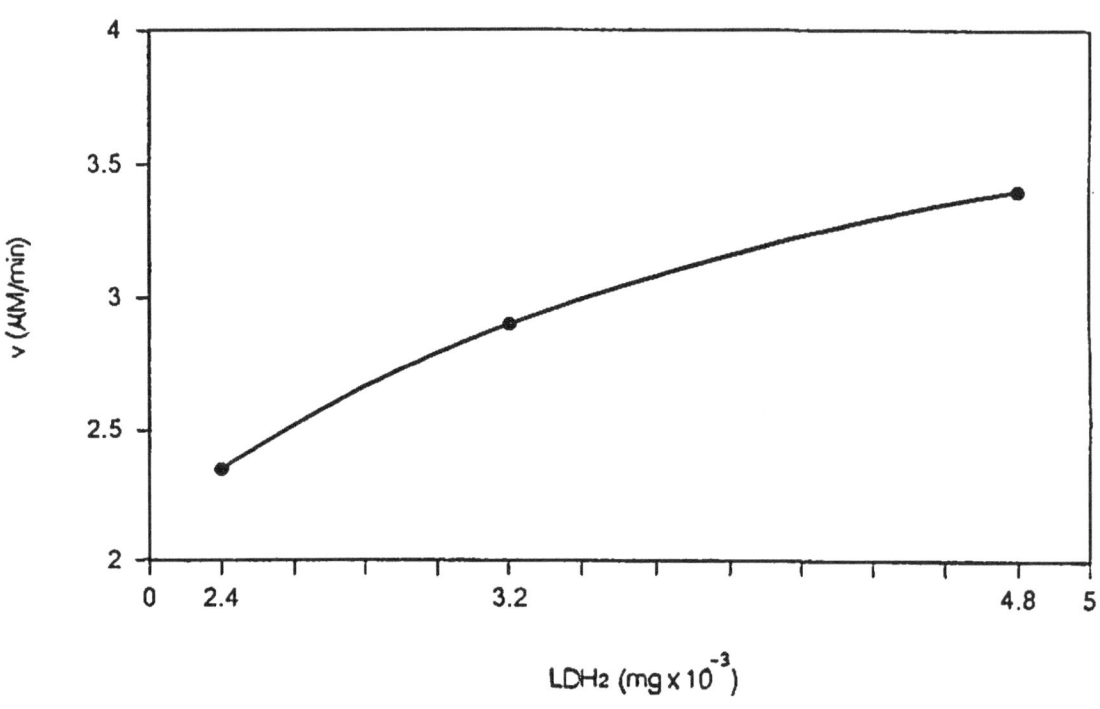

Figure (3.11).    The effect of enzyme concentration on the initial rate.
Details are described in section (2.2.5.5).

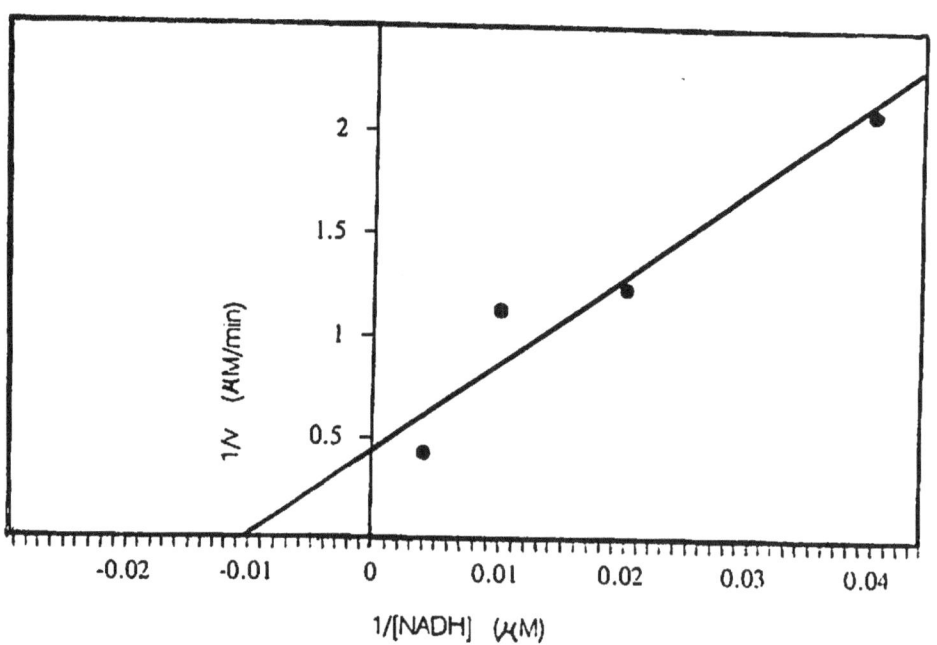

Figure (3.12).    Determination of $K_m$ for NADH using the Lineweaver-Burk plot.

### 3.1.5.6  Reaction order and kinetics of $LDH_2$-NADH complex formation

Time-course data can be tested for conformity to a certain order, by many equations that express reaction rates. The time-course of the formation of the $LDH_2$-NADH complex at four different temperatures are illustrated in figure (3.13 A, B, C, D).

A first order reaction is "one for which at a given temperature, the rate of the reaction depends only on the first power of the concentration of a single reacting species"[107]. The first order rate reaction can be expressed by the following equation :

$$\log C = -\frac{k_1}{2.303} t + \log C_0 \qquad \text{------- (1)}$$

For the $LDH_2$ catalyzed reaction, $C_0$ represents the initial concentration of the substrate NADH, while C represents the concentration of NADH after time ( t ). A reaction is first order if a plot of log C against t gives a straight line with a slope equal to the first order rate constant ( $k_1$ ), divided by the value 2.303. Figure (3.14) shows that the time-course data fit the first order rate equation.

A second order reaction is that where "the rate of the reaction is proportional to the square of the concentration of one of the reagents, or to the product of the concentrations of two species of the reagents" [107]. The same rate equation would be obtained for the two situations if the two species had the same initial concentration :

$$\frac{1}{C} - \frac{1}{C_0} = k_2 t \qquad \text{------- (2)}$$

A plot of 1/C versus t should give a straight line, if the reaction is second order, with a slope equal to the second order rate constant ( $k_2$ ).

Since the two species, NADH and pyruvate, are not equal in their initial concentration, equation (2) cannot be used. When the two species are different in their initial concentration, the following equation could be applied :

$$\frac{1}{a-b} \ln \frac{b(a-x)}{a(b-x)} = K_2 t \qquad \text{------- (3)}$$

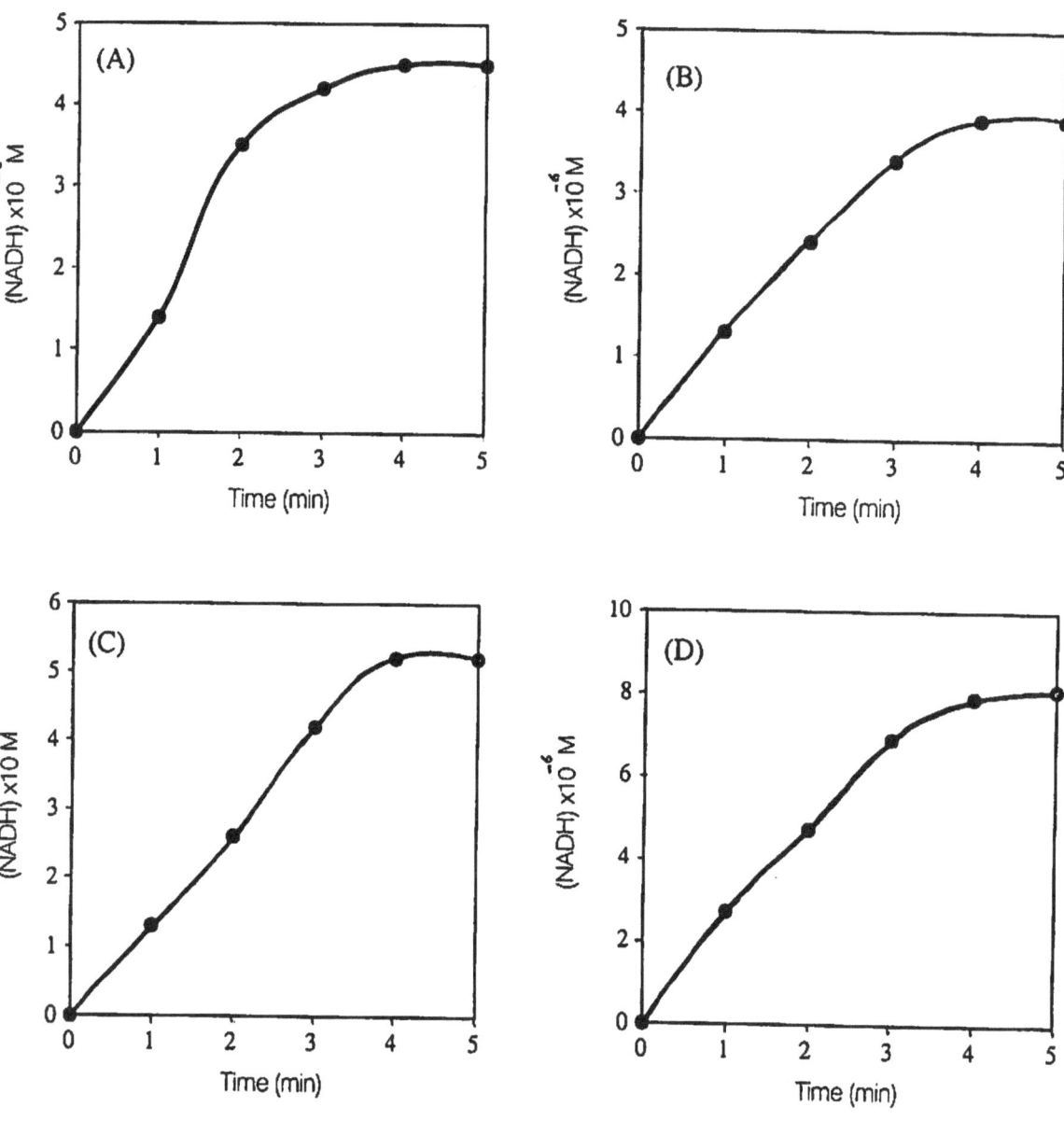

Figure (3.13).  The time-course for the association of  NADH with LDH₂ at 10°C
(A),  25°C (B), 30°C (C),  35°C (D).  Details are described in
section  (2.2.5.6).

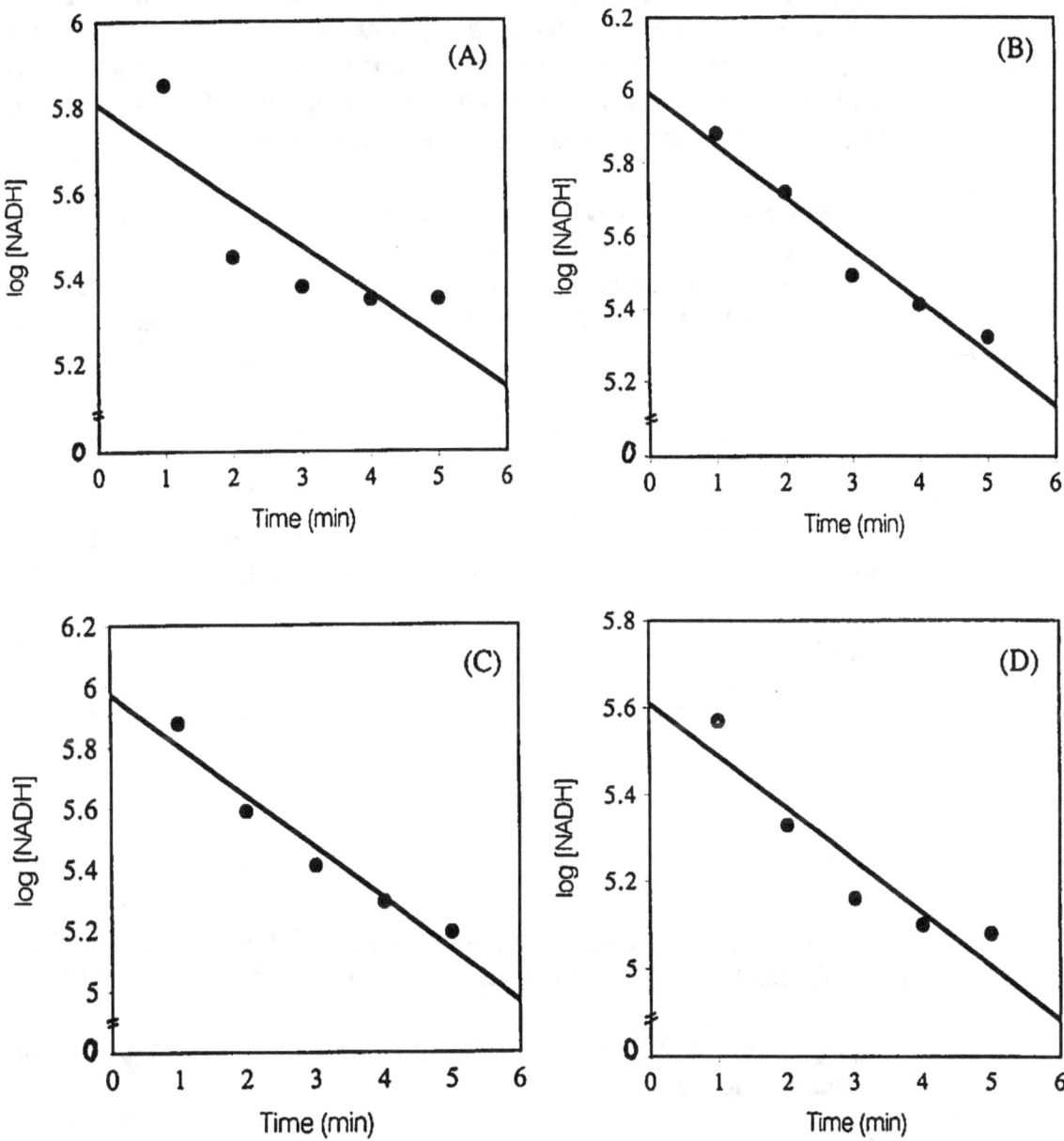

Figure (3.14).    Application of the first order rate equation at 10° C (A), 25° C
(B),30°C  (C), 35° C (D).

Where a, b represent the initial concentration of the two species A, B respectively. A plot of $\{1/ (a-b).\ln b(a-x) / a(b-x)\}$, versus t gives a straight line, if the reaction is second order, with a slope equal to the rate constant ($k_2$). For the $LDH_2$ catalyzed reaction, the two substrates NADH and pyruvate have different initial concentrations and so equation (3) would be convenient to apply. But, the method used in following the reaction depends on the consumption of NADH only, so pyruvate is not followed. To this extent, equation (3) cannot be used.

More comprehensive equations were used to test the conformity of the time course data to a certain order. These equations cover the kinetics of the reaction, specifically the formation of the $LDH_2$-NADH complex, as well as its order. The binding of NADH to $LDH_2$ could be represented by the following reaction :

$$LDH_2 \ + \ NADH \ \underset{k_{-1}}{\overset{k_{+1}}{\rightleftharpoons}} \ LDH_2 \text{ - } NADH$$

Where $k_{+1}$ is the rate constant for the association of NADH with $LDH_2$ and $k_{-1}$ is the rate constant for the dissociation of the complex formed under the same conditions.

At equilibrium :

$$k_a \ = \ [LDH_2 \text{ - } NADH] / [NADH] [LDH_2] \qquad \text{------- (4)}$$

$$k_d \ = \ [NADH] [LDH_2] / [LDH_2 \text{ - } NADH] \qquad \text{------- (5)}$$

$$\therefore \quad k_a \ = \ 1 / k_d \ = \ k_{+1} / k_{-1} \qquad \text{------- (6)}$$

Where $k_a$ is the equilibrium constant of the association (affinity constant) and $k_d$ is the equilibrium constant of dissociation of the $LDH_2$-NADH complex.

The second order rate equation that expresses the above reaction, could be expressed as follows :

$$\ln \overline{X} \{ [S_0 - (\overline{X} X / P_0)] / S_0 (\overline{X} - X) \} \ = \ k_{+1} t [ (S_0 P_0 - \overline{X}^2) / \overline{X}] \qquad \text{------- (8)}$$

Where :
$k_{+1}$ : The kinetic association constant in $M^{-1}. min^{-1}$.
$S_0$ : The total concentration of the substrate NADH in M.
$P_0$ : The total concentration of NADH binding sites on $LDH_2$.
$\overline{X}$ : The concentration of the $LDH_2$-NADH complex at equilibrium.
$X$ : The concentration of the $LDH_2$-NADH complex at time (t).

Equation (8) could be simplified and written as follows :

$$\ln \left[ (\bar{X} - X) / (P_o S_o - \bar{X} X) \right] = k_{+1} t \left[ (P_o S_o - \bar{X}^2) / X \right] - \ln \left[ P_o S_o / \bar{X} \right] \quad \text{------} \quad (9)$$

The concentrations mentioned above were determined from the spectroscopic measurements and by the use of Lambert - Beers law. $\bar{X}$ was calculated from the following equation (108) :

$$\bar{X} = \left[ (P_o + S_o + k_d) / 2 \right] - \left[ (P_o + S_o + k_d) / 2 \right]^2 - P_o S_o \quad \text{------} \quad (10)$$

Both $k_d$ and $P_o$ were calculated from the scatchard plot at four different temperatures as shown in figure (3.15 A, B, C, D) :

$$[S]_b / [S]_f = (-1 / k_d) [S]_b + n [E]_t / k_d \quad \text{------} \quad (11)$$

Where :
$[S]_b$ :   The concentration of bound substrate.
$[S]_f$ :   The concentration of free substrate.
$n$ :   The number of identical and independent substrate binding sites per molecule of enzyme.
$n[E]_t$ :   The total concentration of substrate binding sites = $P_o$.

A plot of $[S]_b / [S]_f$ versus $[S]_b$, gives a straight line with a slope equal to $(1 / k_d)$ and an intercept on the $[S]_b$ axis equal to $(n [E]_t)$. However, in equation (9) $\ln [(\bar{X} - X) / (P_o S_o - \bar{X} X)]$ was plotted as function of time ( t ). A linear plot was obtained at four different temperatures and as shown in figure (3.16 A, B, C, D). This means that the reaction conforms to the rate law of equation (9).

From the concentration measurements, the amount of bound NADH was found to be less than 10% of the total NADH concentration (10% of $7.5 \times 10^{-4}$ M NADH is equal to $7.5 \times 10^{-5}$ M). This means that most of NADH remained free. In other words, the value of $P_o S_o \gg \bar{X}^2$ and $P_o S_o \gg \bar{X} X$. As a result, equation (8) could be written as follows (109) :

$$\ln [\bar{X} / (\bar{X} - X)] = k_{+1} t (P_o S_o / \bar{X}) \quad \text{------} \quad (12)$$

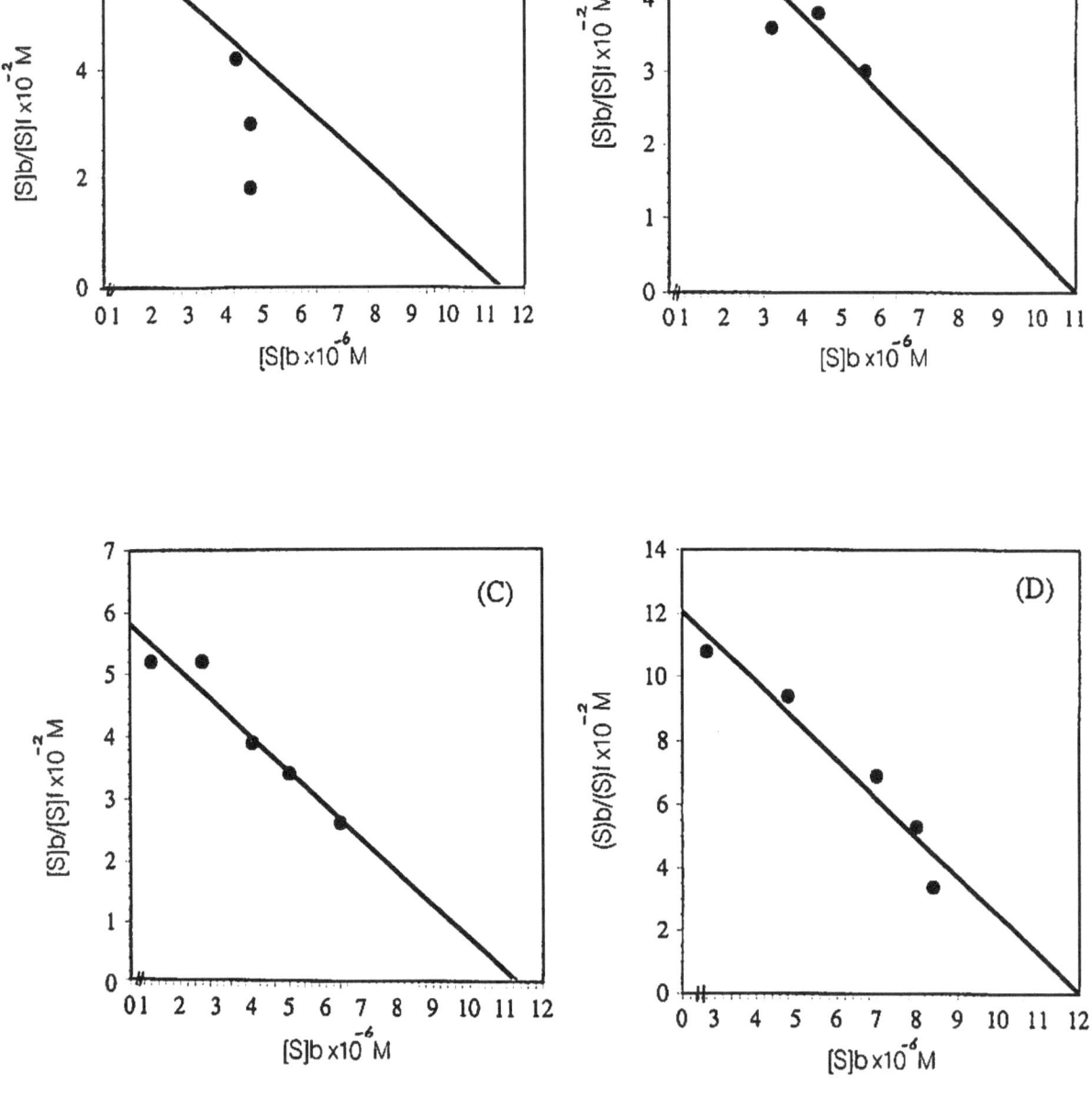

Figure (3.15).  Scatchard Plot of NADH binding to LDH₂ at 10°C (A), 25°C (B), 30°C (C), 35°C (D).  Details are described in section (2.2.5.6).

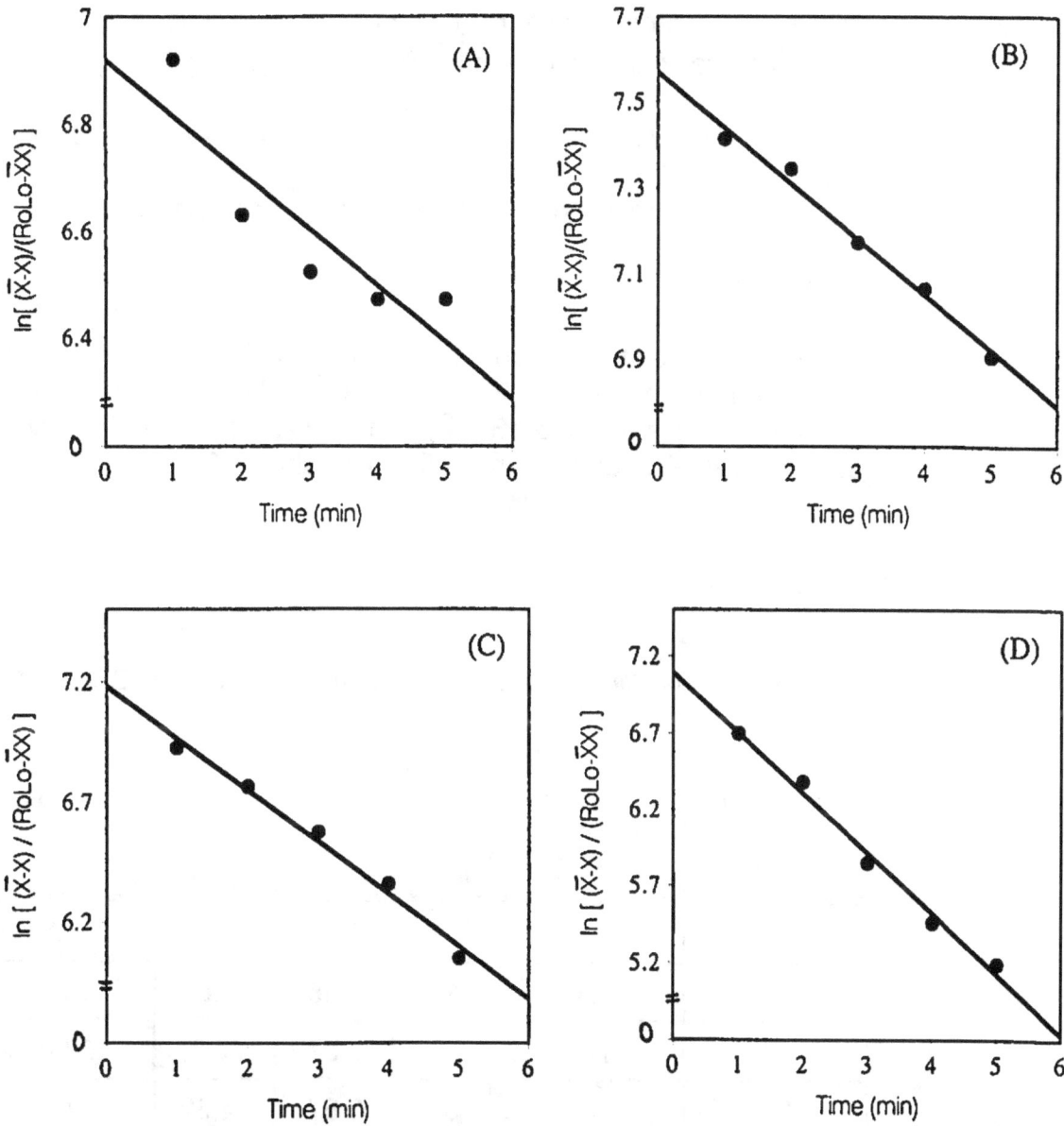

Figure (3.16). Second order kinetics of complex formation between NADH and LDH$_2$ at 10°C (A), 25°C (B), 30°C (C), 35°C (D).

This equation represents a first order rate law. ln $[\overline{X} / (\overline{X} - X)]$ was plotted versus time ( t ) and a straght line was obtained at four different temperatures, as shown in figure (3.17 A,B,C,D). The slope of the straight line represents the observed value of first order rate constant ($k_{obs.}$) in $min^{-1}$. However, the rate constant ($k_{+1}$) in $M^{-1}. min^{-1}$ was calculated from the following equation :

$$k_{obs.} = k_{+1} P_o S_o / \overline{X} \qquad\qquad \text{------- (13)}$$

The dissociation rate constant was calculated from the equation :

$$k_{-1} = k_{+1} \overline{P} \overline{S} / \overline{X} \qquad\qquad \text{------- (14)}$$

Where $\overline{P}$ represents the concentration of NADH's binding sites on $LDH_2$ at equilibrium and is equal to $(P_o - \overline{X})$, while $\overline{S}$ is the concentration of free NADH at equilibrium and is equal to $(S_o - \overline{X})$. The $k_a$ values were calculated from equation (6). All the kinetic parameters are listed in table (3.8).

Table (3.8).  The kinetic parameters for the binding of NADH to $LDH_2$ at different temperatures.

| Temperature (°C) | Binding site concentration x $10^{-6}$ M | $k_{obs.}$ x $10^{-2}$ $min^{-1}$ | $k_{+1}$ x $10^{2}$ $M^{-1}.min^{-1}$ | $k_{-1}$ x $10^{-3}$ $min^{-1}$ | $k_a = k_{+1}/k_{-1}$ x $10^{3}$ $M^{-1}$ | $k_d = k_{-1}/k_{+1}$ x $10^{-5}$ M |
|---|---|---|---|---|---|---|
| 10 | 11.4 | 11 | 1.29 | 13.36 | 9.66 | 10.34 |
| 25 | 11 | 13 | 1.58 | 11.69 | 13.52 | 7.40 |
| 30 | 11.4 | 22 | 2.57 | 25.7 | 10.00 | 10.00 |
| 35 | 12 | 33 | 3.67 | 54.32 | 6.76 | 14.80 |

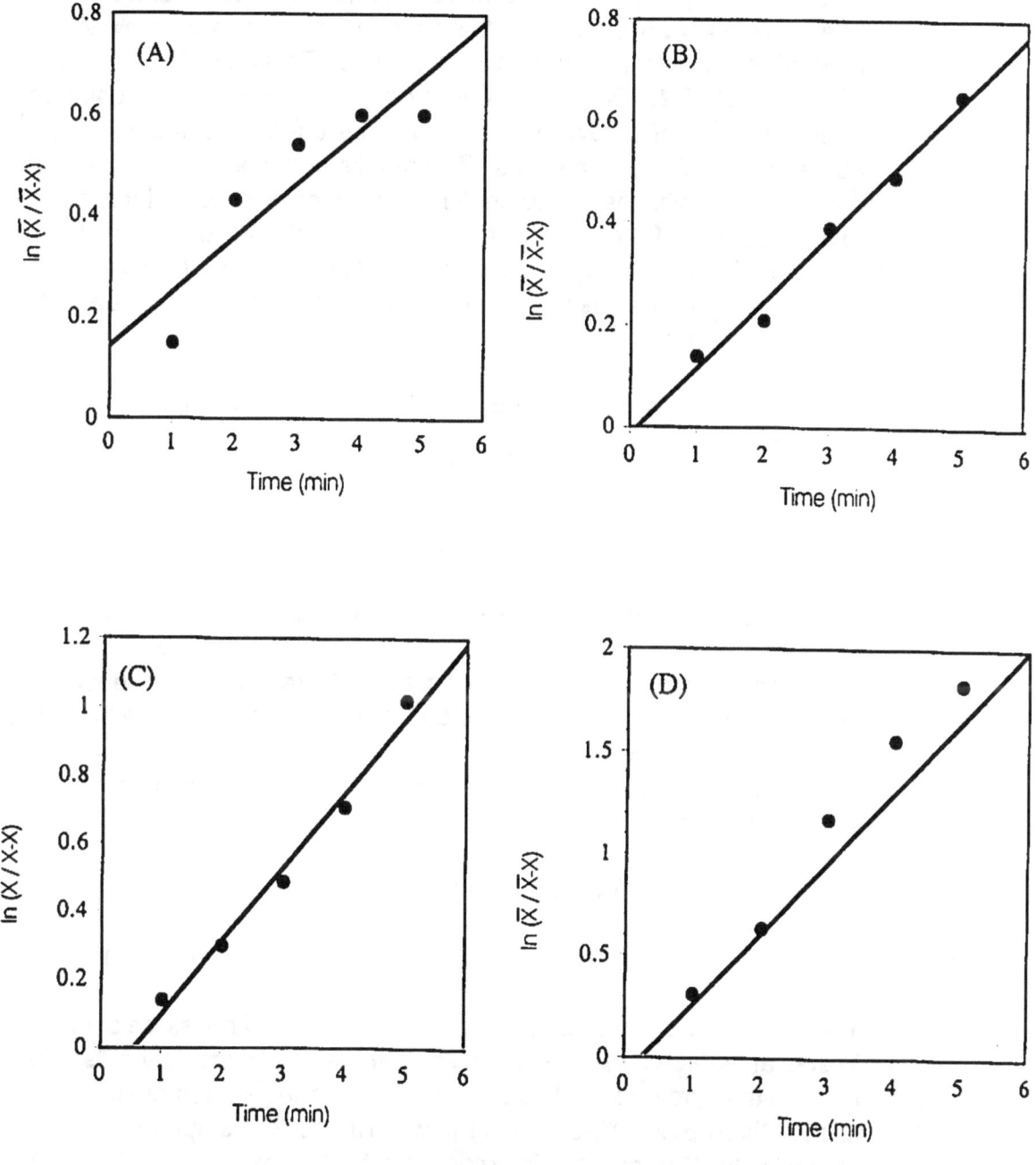

Figure (3.17). First order kinetics of complex formation between NADH and LDH$_2$, at 10°C (A), 25°C (B), 30°C (C), 35°C (D).

### 3.1.5.6 Inhibition of LDH$_2$

**a. Inhibition by urea :** A concentration of 2 M urea was found to have an inhibitory effect on the activity of LDH. At this concentration, the slow moving isoenzymes are more affected than the fast moving ones [78]. Figure (3.18) illustrates the inhibitory effect of (0.1, 0.25, 0.5, 1, 2) M urea on the activity of LDH$_2$, using three different concentrations of pyruvate. The inhibitory effect was found to increase as the concentration of urea increases. The results obtained from figure (3.18), show that the type of inhibition is noncompetitive. A noncompetitive inhibitor binds reversibly to a site different from that which binds the substrate. Both inhibitor and substrate can bind simultaneously to an enzyme molecule without affecting ones another's binding. The mechanism of noncompetitive inhibition could be illustrated by the following scheme :

$$
\begin{array}{ccc}
E + \text{Pyruvate} \rightleftharpoons & E\text{-Pyruvate} \longrightarrow E + P \\
+ & + \\
\text{Urea} & \text{Urea} \\
\\
k_i \updownarrow & k_i \updownarrow \\
\\
E\text{-Urea} + \text{Pyruvate} \rightleftharpoons & E - \text{Pyruvate - Urea}
\end{array}
$$

The noncompetitive inhibitor acts by decreasing the turnover number rather than by reducing the proportion of enzyme molecules that are bound to substrate. However, noncompetitive inhibition cannot be overcome by increasing the substrate concentration. The velocity equation for this type of inhibition could be represented by the Dixon equation:

$$
\frac{1}{V} = \frac{1 + k_m / [\text{pyruvate}]}{V_{max}\ k_i} [\text{Urea}] + \frac{1}{V_{max}} \left(1 + \frac{K_m}{[\text{Pyruvate}]}\right)
$$

The plot of $1/V$ versus [Urea], gives a straight line with a slope of $(1 + k_m/[\text{pyruvate}])/V_{max}\ k_i$, an intercept of $1/V_{max}(1 + km/[\text{pyruvate}])$ on the $1/V$ axis and an intercept of $-k_i$ on the [Urea] axis. Table (3.9) illustrates the values of slopes and intercepts of the Dixon plot. At (5, 2, 0.7) mM pyruvate, the slopes were 0.02, 0.03, 0.04 respectively. Whereas the intercepts on the $1/V$ axis were 0.025, 0.03, 0.06 at (5, 2, 0.7) mM pyruvate respectively. The value of $k_i$ was found to be 1.1 M. A replot of the slopes obtained from Dixon plot were plotted versus $1/[\text{pyruvate}]$ and a straight line was obtained, with a slope of $k_m/V_{max}\ k_i$, an intercept of $1/V_{max}$ on the (slope) axis and an intercept of $-1/k_m$ on the $1/[\text{pyruvate}]$ (figure 3.19). From this plot, the values of $V_{max}$ and $k_m$ were determined. $V_{max}$ was found to be 48 U/L, while $k_m$ was 0.8 mM.

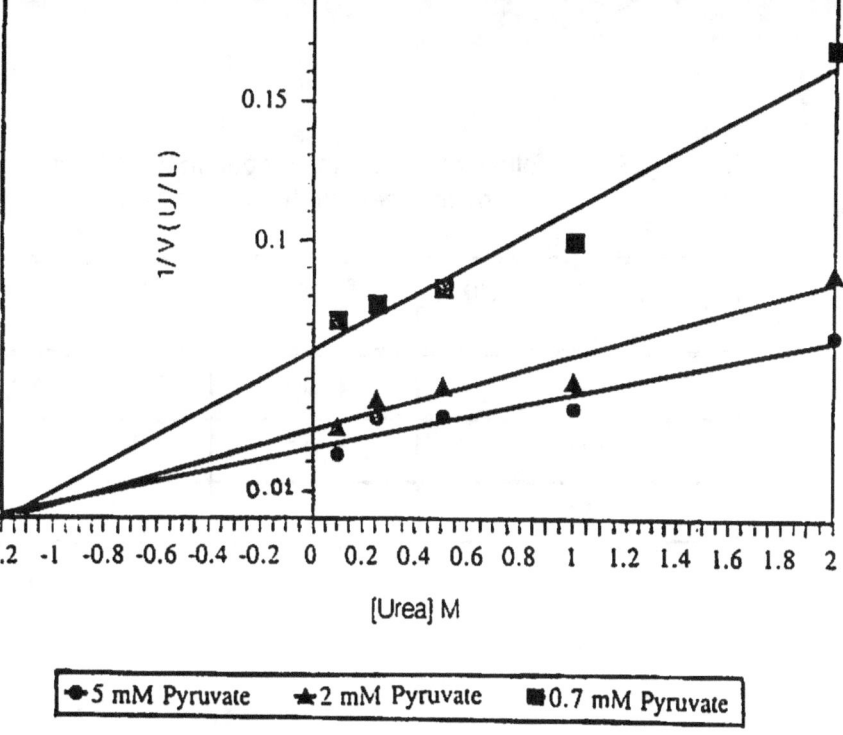

Figure (3.18). Dixon plot for the inhibition of LDH₂ by (0.1, 0.25, 0.5, 1, 2) M urea using (0.7, 2, 5) mM pyruvate.The plot indicates that LDH 2 is noncompetetively inhibited by urea. Details are described in section (2.2.5.7).

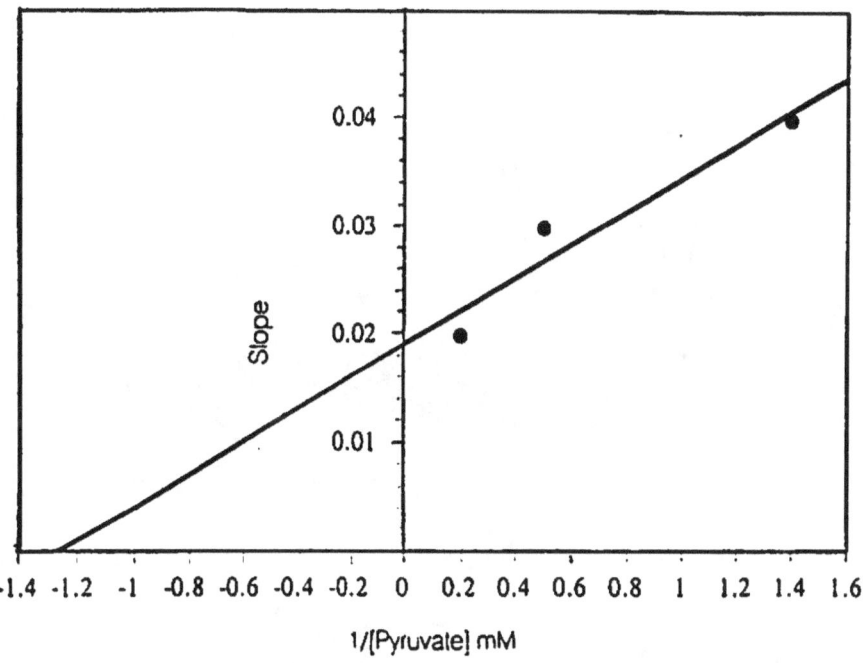

Figure (3.19). Replot of the slopes of Dixon plot of the inhibition of LDH₂ by urea.

Table (3.9).   Values of slopes, intercepts on the 1/ V axis (1/ Vmax ( 1+ km/[pyruvate] ))
and the intercept on the [urea] axis (ki) obtained from Dixon plot.

| [Pyruvate] (mM) | Slope | 1/ $V_{max}$ (1+$k_m$ / [pyruvate]) | $k_i$ (M) |
|---|---|---|---|
| 5 | 0.02 | 0.025 | 1.1 |
| 2 | 0.03 | 0.03 | 1.1 |
| 0.7 | 0.04 | 0.06 | 1.1 |

**b. Inhibition by oxalate :** Oxalate is known to have an inhibitory effect on LDH.  At a concentration of 0.2 mM oxalate, the fast moving isoenzymes are more affected than the slow moving ones [87].  Figure (3.20) shows the inhibitory effect of (0.25 ,0.125, 0.062, 0.031, 0.031) mM oxalate on the activity of $LDH_2$, at three different pyruvate concentrations.  The results obtained from the figure indicate that the type of inhibition does not conform to any of the known types of inhibition.

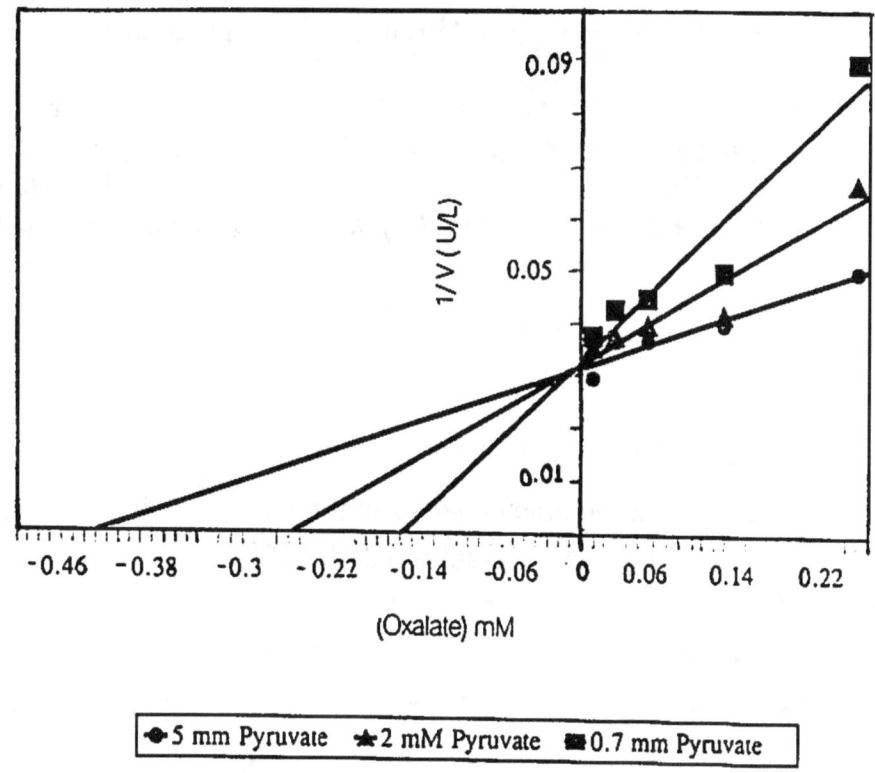

Figure (3.20). Dixon plot for the inhibition of LDH₂ by (0.25, 0.125, 0.062, 0.031, 0.01) mM oxalate using (0.7, 2, 5) mM pyruvate. The plot indicates that the type of inhibition does not appyl to any of the known types of inhibition. Details are described in section (2.2.5.7).

### 3.1.6 The thermodynamics of NADH binding to LDH$_2$

**a. The thermodynamic parameters of the standard state :**

The effect of temperature in the affinity constant ( $k_d$ ), was studied through Van't Hoff's plot (figure 3.21). The values of log $k_a$ obtained from scatchard plots at different temperatures, were plotted versus the reciprocal values of absolute temperatures in Kelvin (1/ T), according to the following equation :

$$\ln k_a = \Delta S° / R - \Delta H° / RT$$

Where :

$\Delta S°$ : The entropy change of the standard state.

$\Delta H°$ : The enthalpy change of the standard state.

R   : The gas constant = 8.314 J. K$^{-1}$.

$\Delta H°$was obtained from the slope of the plot. The change in Gibb's free energy of the standard state ($\Delta G°$) was obtained from the following equation :

$$\Delta G° = - RT \ln k_a$$

$\Delta S°$ was obtained from the following equation :

$$\Delta S° = (\Delta H° - \Delta G°) / T$$

The values of the thermodynamic parameters of the standard state are listed in table (3.10).

Table (3.10). The thermodynamic parameters for the binding of NADH to LDH$_2$ at 35°C. Details are described in section (2.2.6).

| $\Delta H°$ KJ.mol$^{-1}$ | $\Delta G°$ KJ.mol$^{-1}$ | $\Delta S°$ J.deg$^{-1}$.mol$^{-1}$ | $\Delta H^*$ KJ.mol$^{-1}$ | $\Delta G^*$ KJ.mol$^{-1}$ | $\Delta S^*$ J.deg$^{-1}$.mol$^{-1}$ | E$_a$ KJ.mol$^{-1}$ |
|---|---|---|---|---|---|---|
| 17.5 | -24.2 | 135.4 | 5.42 | 65.6 | -195.4 | 7.981 |

The results indicated that $\Delta H°$ had a positive value, which means that the reaction was endothermic. The negative value of $\Delta G°$ indicated that the reaction was spontaneous and that the overall reaction was energetically favorable in the direction of complex formation. The positive value of $\Delta S°$ indicated that the entropy was the driving force for the reaction. In other words, the high negative value of $\Delta G°$ was controlled by the high positive value of $\Delta G°$. This refers to the stability of the complex and the high affinity of the reactants.

The positive $\Delta H°$ value may refer to favorable interactions between groups within both $LDH_2$ and NADH. These include the non covalent interactions which are electrostatic in nature. Such as charge-charge interactions, charge-dipole, dipole-dipole and hydrogen bonds. These interactions all together help in stabilizing the folded structure of the complex. This is conformed by the large positive entropy change (110).

## b. The thermodynamic parameters of the transition state

The thermodynamic parameters of the transition state were studied through the Arrhenius equation, this equation relates the logarithm of the rate constant k or $V_{max}$ (in biochemical reactions) with the reciprocal values of absolute temperatures in Kelvin (1/T), according to the following formula :

$$\log V_{max} = \log A - E_a / 2.303\, R \,.\, 1/T$$

Where A is the Arrhenius constant and $E_a$ is the energy of activation. A plot of log $V_{max}$ versus 1/T gives a straight line with a slope equal to the value of $E_a$ (figure 3.22). The enthalpy change of the transition state ($\Delta H^*$) was obtained from the following equation :

$$\Delta H^* = E_a - RT$$

While the Gibb's free energy of the transition state ($\Delta G^*$) was obtained from the following equation :

$$\Delta G^* = -RT \ln V_{max} + RT \ln (kT / h)$$

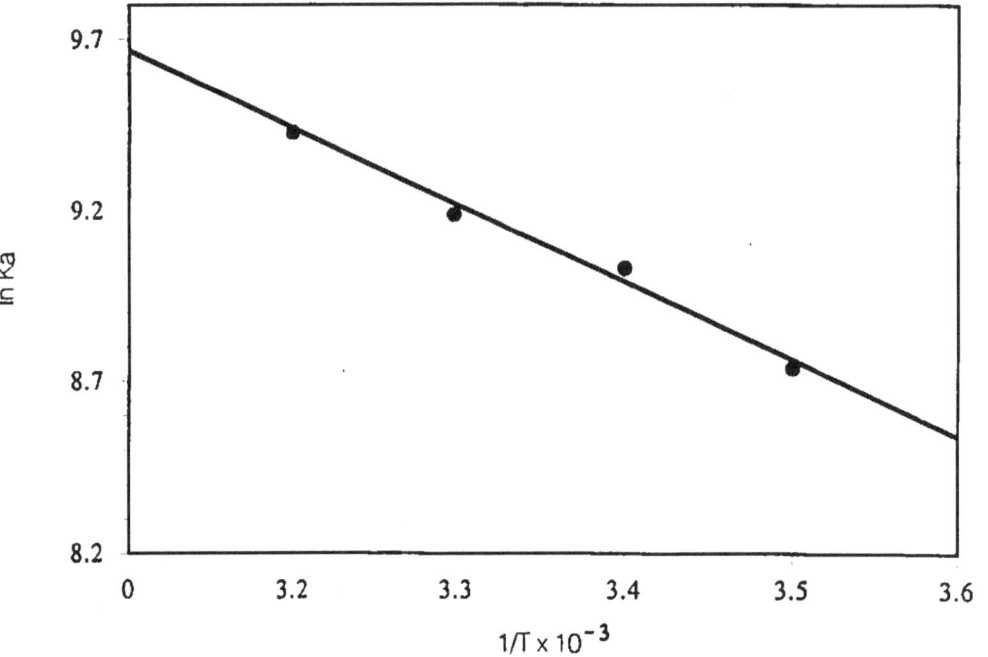

Figure (3.21). Van't Hoff's Plot (the dependance of the equilibrium binding constant (ka) on the temperature) for the NADH binding to $LDH_2$. Details are described in section (2.2.6).

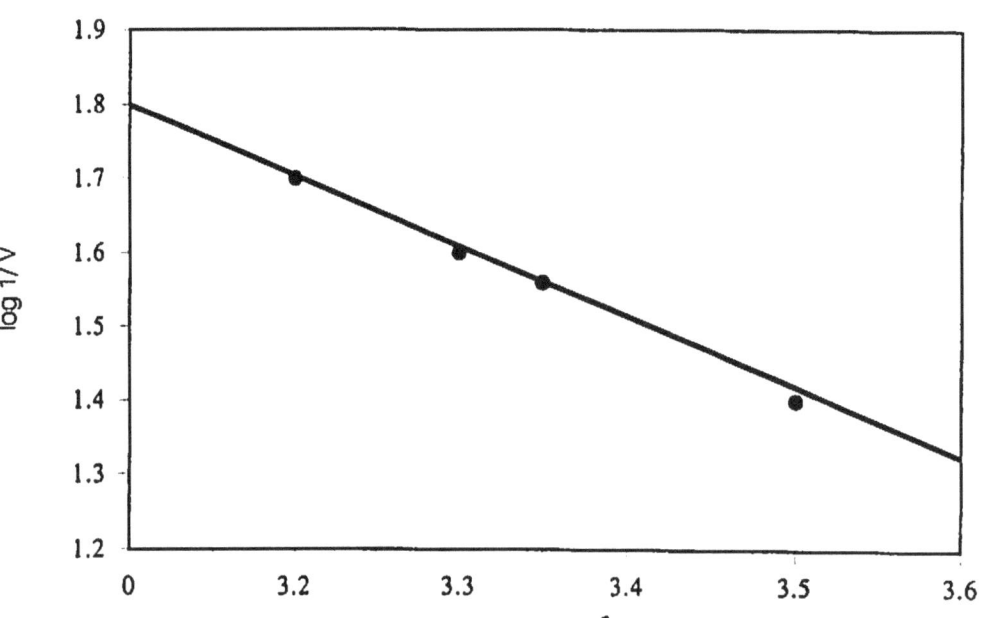

Figure (3.22). Arrhenius plot for the NADH bonding to $LDH_2$. Details are described in section (2.2.6).

Where (k) represents the Boltzmann constant and (h) Plank's constant. The entropy change of the transition state ($\Delta S^*$) was obtained from the following equation :

$$\Delta S^* = (\Delta H^* - \Delta G^*) / T$$

The values of the thermodynamic parameters of the transition state are listed in table (3.10).

The positive value of $\Delta H^*$ for the formation of the LDH$_2$-NADH complex, indicated that the reaction is endothermic. The high positive value of $\Delta G^*$ indicated that the formation of the activated complex was nonspontaneous and required a lot of energy to overcome the transition state (energy barrier), leading to the final product. $\Delta S^*$ had a negative value, which means that the activated complex had a more ordered structure than the reactants. The positive value of $\Delta G^*$ is mainly attributed to the decrease in entropy of the transition state ($\Delta S < 0$).

The information obtained from the thermodynamic parameters, gave an overall idea about the nature of forces that control the complex formation. From this information, a thermodynamic model describing the complex formation was suggested. The model is illustrated in figure (3.23).

The model proposes that the formation of the LDH$_2$-NADH complex undergoes three thermodynamic states. The thermodynamic state A represents the initial energy level of LDH$_2$ and NADH. In the thermodynamic state B, the two species bind to form the activated complex [LDH$_2$-NADH]. The last thermodynamic state C, represents the formation of the fully interacting complex LDH$_2$-NADH. In step 1 of the reaction, the binding of NADH to LDH$_2$ was associated with a positive $\Delta G^*$ value. This means that the initial step of the reaction requires an input of energy for the system. The negative entropy change for this step indicates that the activated complex has a more ordered structure than the reactants. In step 2, the activated complex participates in further interactions, giving the fully interacting complex LDH$_2$-NADH. It is proposed that the formation of a protein-ligand complex, occurs in two steps. The first is the stabilization of the complex by hydrophobic interactions, and the second is the stabilization by short range interactions, such as electrostatic interactions, hydrogen bonding and Van der Waals interactions [111]. Hydrophobic interactions contribute to the complex stability via high positive entropy changes ($\Delta S > 0$), while electrostatic interactions, hydrogen bonding and Van der

Waals interactions contribute to the stability of the complex via negative changes in standard enthalpy change [111], [112].

The thermodynamic data from our study indicated that the binding of NADH to $LDH_2$ are entropy driven and come in agreement with the concept that hydrophobic interactions play an important role in NADH $LDH_2$ interactions.

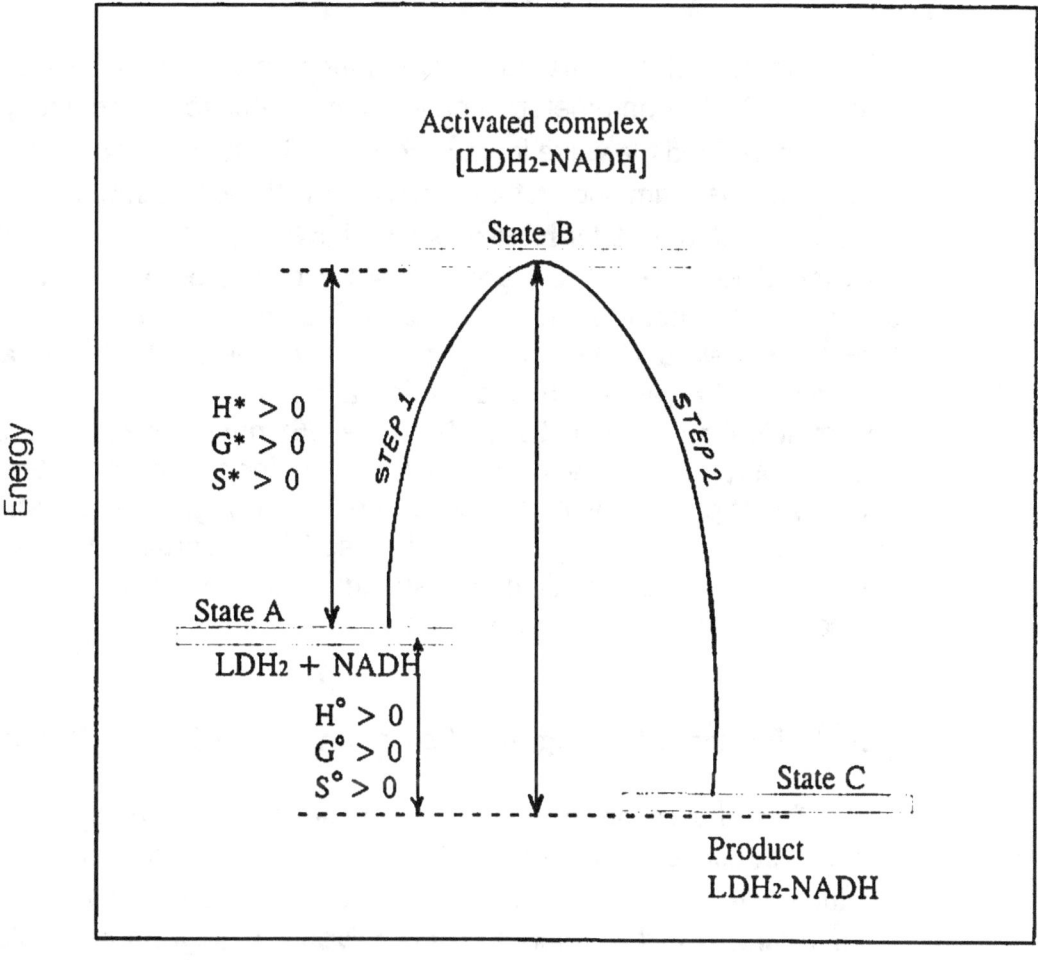

Figure (3.23).    General energy diagram and thermodynamic model applied to the complex formation between LDH₂ and NADH.

### 3.1.7 Spectroscopic studies on LDH$_2$ and LDH$_5$

#### 3.1.7.1 The U.V spectrum of LDH$_2$ and LDH$_5$

Figure (3.24) illustrates the U.V spectrum of both LDH$_2$ and LDH$_5$ at pH 8.0 and at 30°C. The parameters measured are the absorbance, the molar absorption coefficient ($\epsilon$) and the wave length corresponding to a peak of maximum absorption ($\lambda_{max}$). The spectrum shows that the $\lambda_{max}$ for LDH$_2$ is 220 nm, while the highest $\lambda_{max}$ for LDH$_5$ is 268 nm. $\epsilon$ for LDH$_2$ was found to be 13.06 L / gm.cm, while $\epsilon$ for LDH$_5$ was found to be 11.21 L / gm.cm. As a result, both LDH$_2$ and LDH$_5$ have a characteristic spectrum and can be identified by the peaks in each spectrum. For LDH$_2$, the peak at 220 nm is assigned to tyrosine . The two peaks in the LDH$_5$ spectrum, at 220 nm are also assigned to tyrosine. Hence, LDH$_2$ and LDH$_5$ both absorb at 220 nm but only LDH$_5$ absorbs at 268 nm. This could be attributed to the structure of tyrosine in each isoenzyme. It seems that in LDH$_2$, tyrosine is located in a way that part of it is on the surface of the protein molecule and the other part is buried, so not all of tyrosine is exposed to absorbance. Whereas in LDH$_5$, tyrosine seems to be totally on the surface and all of it is exposed to absorbance. Hence, LDH$_5$ showed two tyrosine peaks.

#### 3.1.7.2 Factors affecting the absorption properties of LDH$_2$ and LDH$_5$

The absorption spectrum of any protein is primarily determined by the chemical structure of the protein. However, a large number of environmental factors produce detectable changes in $\lambda_{max}$ and $\epsilon$. The environmental factors consist of temperature, pH, the polarity of the solvent or neighboring molecules and the relative orientation of neighboring chromophores.

#### a. Temperature effect on the U.V spectrum of LDH$_2$ and LDH$_5$

The effect of temperature on both LDH$_2$ and LDH$_5$ are shown in figure (3.25). The absorption was measured at 45°C. The LDH$_2$ spectrum shows that the temperature did not affect the region at 220 nm. Whereas a new $\lambda_{max}$ was found at 274 nm, this wave length is assigned to tyrosine. Mainly, the temperature has an effect on the weak interactions in the protein molecule, such as hydrogen bonds. High temperature may disrupt these interactions. Hence, the appearance of the peak at 274 nm could be attributed to this disruption, which might have caused tyrosine to be more exposed to absorbance. The LDH$_5$ spectrum shows that $\lambda_{max}$ was affected by the temperature. At 30°C (figure 3.24) two peaks were found at 220 nm and 268 nm, the highest $\lambda_{max}$ was at 368 nm. At 45°C the same peaks were found, but this time the highest $\lambda_{max}$ was at 220 nm. The two peaks, as previously mentioned, belongs to tyrosine. However, additional peaks were also found at 244 nm and 284 nm assigning to cysteine and tryptophane respectively.

### b. pH effect on the U.V spectrum of LDH₂ and LDH₅

The pH of the solvent determines the ionization state of both LDH$_2$ and LDH$_5$. Figure (3.26) shows the spectrum of both LDH$_2$ and LDH$_5$ at pH 7.0 and 30°C. The LDH$_2$ spectrum showed three additional peaks; at 246, 260 and 278 nm. These peaks are assigned to cysteine, phenylalanine and tryptophane, respectively. The LDH$_5$ spectrum, showed a disappearance of the peak at 268 nm and an additional peak at 226, which is assigned to tyrosine. The amino acids that appeared in the LDH$_2$ spectrum at pH 7.0, did not appear at pH 8.0. This could be attributed to the effect of neighboring groups which might prevent these amino acids from being exposed to absorbance. The disappearance of these amino acids at pH 8.0 is not attributed to the ionization effect, because these three amino acids have a pK above eight. For the same reasons, the peak representing tyrosine at 268 nm in the LDH$_5$ spectrum disappeared at pH 7.0.

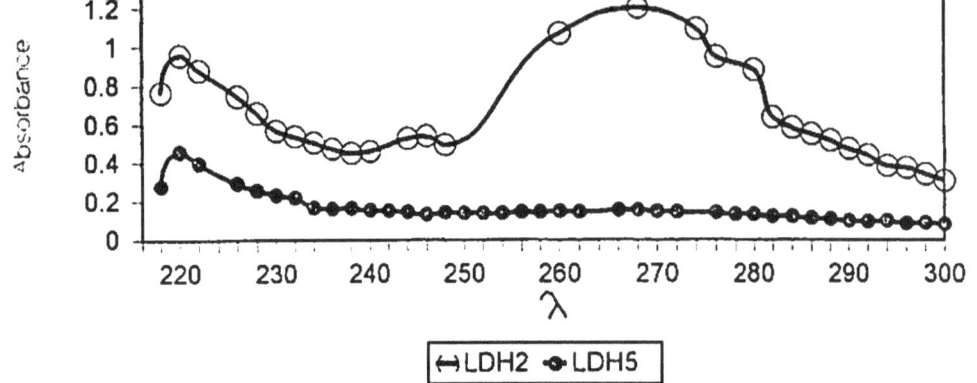

Figure (3.24).   The U.V spectrum of LDH₂ and LDH₅ at pH 8.0
                 and  30°C.  Details are described in section
                 (2.2.7.1).

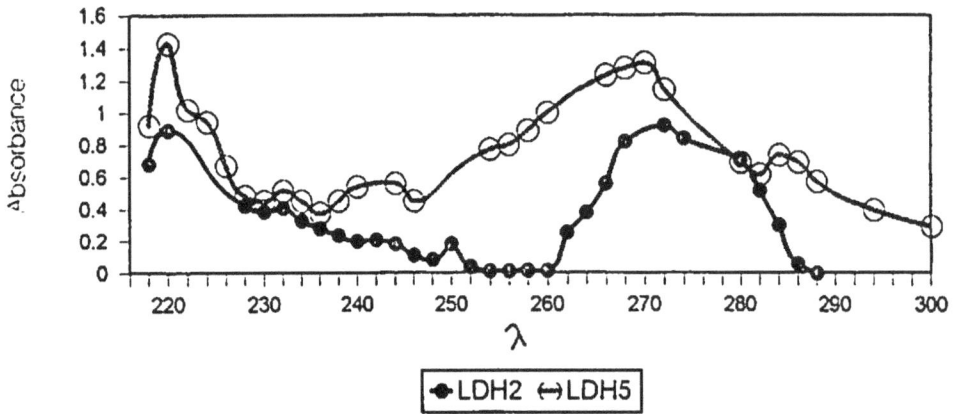

Figure (3.25).   The effect of temperature (45°C) on the U.V spectrum
                 of   LDH₂ and LDH₅. Details are described in section
                 (2.2.7.2).

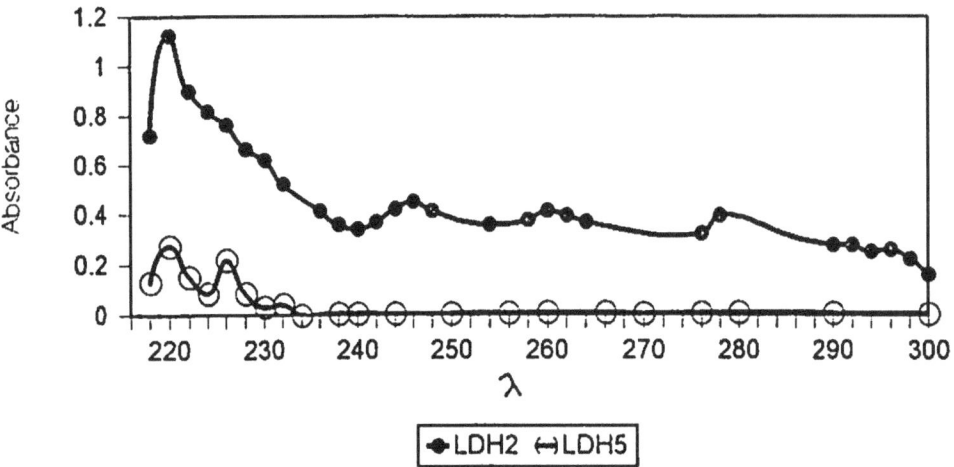

Figure (3.26).   The effect of pH (pH 7.0) on the U.V spectrum of  LDH
                 and LDH₅.  Details are described in section (2.2.7.2).

### 3.1.7.3 Binding studies on LDHₛ with NADH

It is known that the binding of an enzyme with certain compounds produces spectral changes in the chromophores of the enzyme. In order to study these effects, the isoenzyme LDHₛ was selected as a model, because its spectrum was studied. NADH was selected because this compound has a well known spectrum in literature (113). NADH absorbs at two $\lambda_{max}$, one at 260 nm and the other at 340 nm. However, the U.V spectrum of NADH was studied and the results were consistent with the literature; two peaks at $\lambda_{max}$ 260 nm and 340 nm were obtained.

### a. The U.V spectrum of the LDHₛ-NADH complex

Figure (3.27) represents the spectrum of the LDHₛ- NADH complex at pH 8.0 and 30°C. The spectrum shows that $\lambda_{max}$ for NADH was not affected, while $\lambda_{max}$ for LDHₛ was. LDHₛ had a maximum absorbance at 220 nm and 268 nm, whereas the complex showed a maximum absorbance at 260 nm and 340 nm, which is the same $\lambda_{max}$ for NADH. However, the absorbance at 260 nm could be assigned to phenyl alanine. It seems that NADH predominates over LDHₛ, this could be attributed to many reasons. For example, the concentration of NADH is higher than LDHₛ. Another reason could be that the binding of NADH was not at the tyrosine and tryptophane region, because the absorption at 260 nm represents phenyl alanine.

### b. pH effect on the U.V spectrum of the LDHₛ-NADH complex

Figure (3.28) shows the effect of pH 6.0, 6.5, 7.0, 7.5 on the LDHₛ-NADH spectrum. The results indicated that $\lambda_{max}$ was not affected by the different pH values.

### c. LDH5 : NADH concentration effect on U.V spectrum of the LDH5-NADH complex

Figure (3.29) illustrates the effect of using different amounts of LDHₛ and NADH. (0.25, 0.5, 0.75, 1) mL of the fraction containing LDHₛ was added to (2.25, 2, 1.75, 1.5) mL of NADH, respectively. The absorbance was measured at pH 6.5 and 30°C. The results indicated that using different amounts of LDHₛ and NADH had no effect on $\lambda_{max}$.

### b. The stability of the LDHₛ-NADH complex

The absorbance of the LDHₛ-NADH complex was measured at 30°C throughout 50 days. The spectra obtained from these measurements are illustrated in figure (3.30). The spectra show that within 30 days, the phenyl alanine region ($\lambda_{max}$ at 260 nm) was not affected. However, the NADH region at 340 nm disappeared after 50 days.

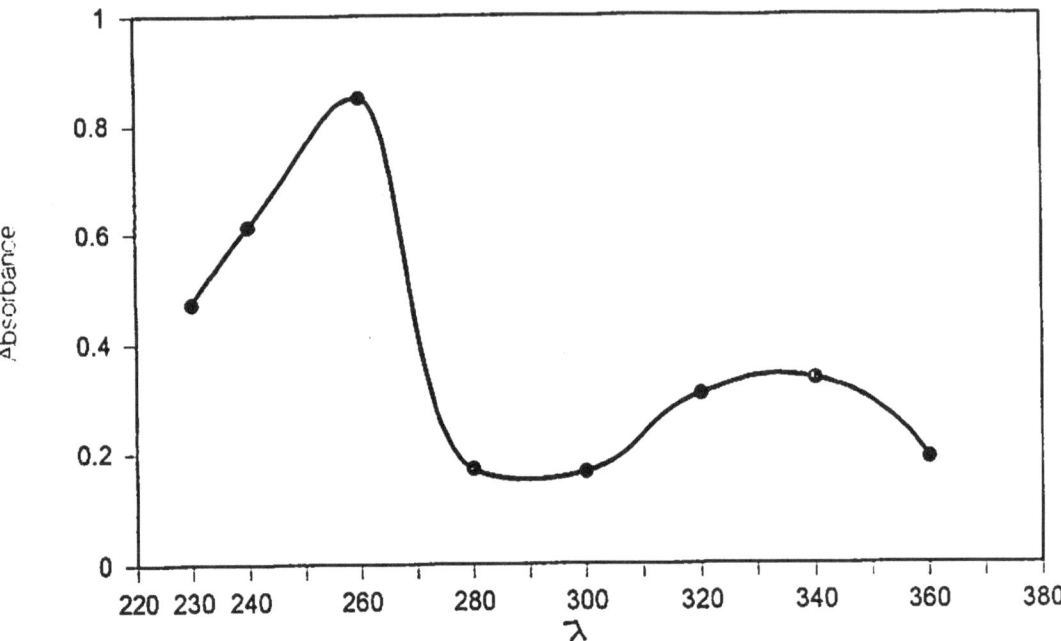

Figure (3.27).   The U.V spectrum of the LDHs-NADH complex at pH 8.0 and  30°C.  Details are described in section (2.2.7.3).

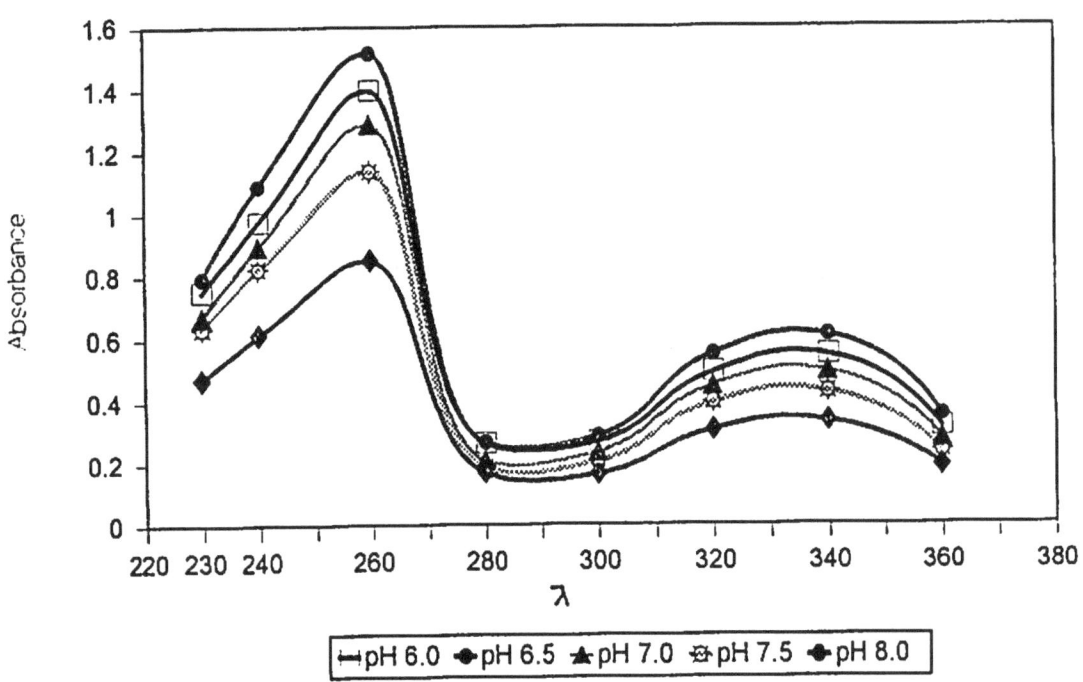

$\vdash$pH 6.0  ◆pH 6.5  ▲pH 7.0  ✿pH 7.5  ◆pH 8.0

Figure (3.28).   The effect of pH on the U.V spectrum of the LDHs-NADH omplex at 30° C.  Details are described in section (2.2.7.3).

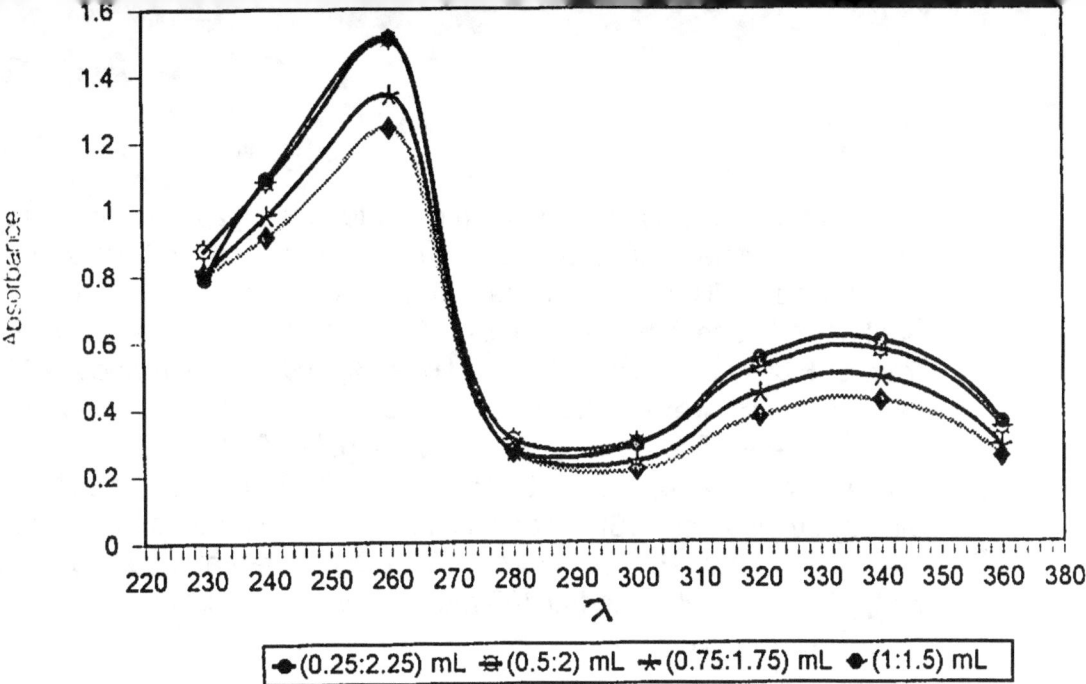

Figure (3.29). LDH₅-NADH concentration effect on the U.V spectrum of the LDH₅-NADH complex, at pH 6.5 and 30°C. Details are described n section (2.2.7.3).

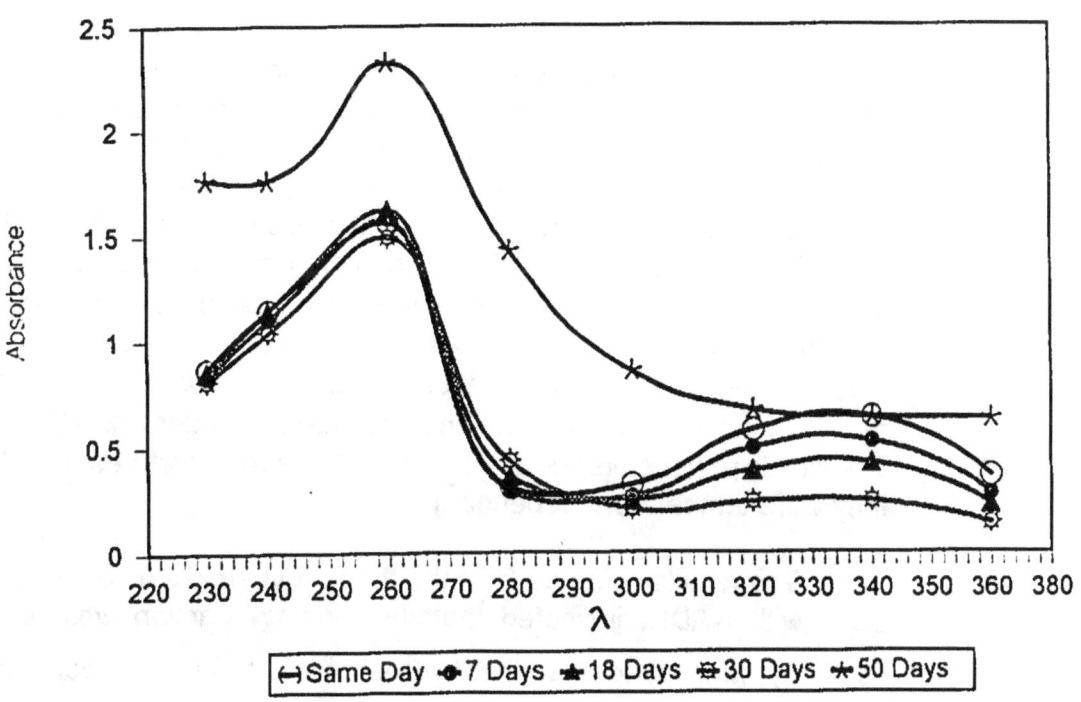

Figure (3.30). The stability of the LDH₅-NADH complex. The complex was stored at 4°C for 50 days and measured at 30°C throughout the days. Details are described in section (2.2.7.3).

## Conclusion

This work was carried out in order to investigate the characteristics of LDH in CSF in some CNS diseases. Activity and total protein concentration were determined in all CSF specimens and were found to rise mostly in the diseases that are known to affect the permeability of the blood-CSF barrier. LDH was separated from CSF by gel filtration with a yield of 98%, and its isoenzymes $LDH_1$, $LDH_2$, $LDH_3$, $LDH_4$ and $LDH_5$ were further separated by anion exchange chromatography using a pH gradient. The isoenzymes eluted at pH (7.0, 7.5, 7.8, 8.1, 8.9) respectively, with a yield of (14, 4, 3, 2, 7)%, respectively. The isoenzymes $LDH_2$ and $LDH_5$ were directly purified from CSF by anion exchange chromatography, without the gel filter step, using (100, 150, 200) mM NaCl. $LDH_2$ eluted at a salt concentration of 200 mM, whereas $LDH_5$ eluted at 100 mM. $LDH_2$ and $LDH_5$ were purified with a yield of (6, 5)% respectively.

The electrophoretic pattern of CSF LDH in some neurological diseases, showed the presence of the five isoenzymes with different amounts, depending on the type of disease. However, the electrophoretic pattern of a bacterial meningitis case revealed an extra band, which was designated $LDH_6$. This extra isoenzyme moved faster than $LDH_1$, which means that its charge is more negative than $LDH_1$.

The kinetic studies on $LDH_2$, showed that this isoenzymes obeys Michaelis-Menton equation. It also showed that substrate concentration, pH and temperature had an effect on the rate of the reaction. From the plots of the velocity of reaction versus each of these factors, the optimum pyruvate concentration for the $LDH_2$ catalyzed reaction was found to be 5 mM, while that of NADH was found to be 0.75 mM. The pH optimum was found to be 7.0, and the optimum temperature was 35°C. The $k_m$ value for NADH was determined using the Lineweaver-Burk plot, and was found to be 0.1 mM. Many equations were applied to test the order of the reaction and to follow the kinetics of the association of $LDH_2$ and NADH. The results indicated that the reaction is pseudo first order and at (10, 25, 30, 35)°C. The kinetic parameters were obtained at these temperatures, and it was found that they were temperature dependant.

The results obtained from the thermodynamic studies on the association of $LDH_2$ with NADH, indicated that the binding reaction was entropically driven ($\Delta S° > 0$) suggesting that the hydrophobic interactions were important in the formation of the $LDH_2$-NADH complex.

The spectroscopic studies on $LDH_2$ and $LDH_5$ revealed a characteristic spectrum for both isoenzymes, at pH 8.0 and 30°C. The maximum absorbance for

LDH$_2$ was at 220 nm, while that for LDH$_5$ was at 220 nm and 268 nm assigning to the amino acid tyrosine. The study also showed that both temperature and pH had an effect on the spectra. The binding of LDH$_5$ with NADH was studied, and a characteristic spectrum of the LDH$_5$-NADH was obtained at pH 8.0 and 30°C. The spectrum showed a maximum absorbance at 260 nm and 340 nm. These two wave lengths represent the maximum absorbance of NADH, indicating that only the spectrum of LDH$_5$ was affected by the binding, and that NADH predominated over LDH$_5$. The study showed that using different pH and different LDH5 : NADH concentration, had no effect on the maximum absorbance.

# REFERENCES

1.  Kaplan, A. (1989), Clinical chemistry, theory, analysis and correlation, 2nd edition, pp 594-606.
2.  Wilkinson, J. (1986), Neuroanatomy for medical students, p. 12.
3.  Reymond, A. (1981), Principles of neurology, 2nd edition, p. 617.
4.  Peele, T. (1977), The neuroanatomic basis for clinical neurology, 3rd edition, pp 62-67.
5.  Oehmichen, M. (1976), Cerebrospinal fluid cytology, p. 16.
6.  Glasser, L. (1981), Diagn Med , Jan-Feb; 23-33.
7.  Plum, F. & Posner J. (1980), The diagnosis of stupor and coma, 3rd edition, pp 40-45.
8.  Fishman, R. (1980), Cerebrospinal fluid in the diseases of the central nervous system, pp 35-47.
9.  Castleberry, R., Moreno, H. & Wallace, L. (1975), J. Pediatr., 86 ; 990-993.
10. Lyons, R. & Andriole V. (1986), Neurol. Clin., 4 ; 159-170.
11. Shamma, A. (1971), Lectures in pathology, pp 354-357.
12. Moyer, K. (1980), Neuroanatomy, pp 36-41.
13. Hammock, M. & Milhorat, T. (1976), Ann. Clin. Lab. Sci., 6 ; 22-26.
14. Reymond, A. (1981), Principles of neurology, 2nd edition pp 429-438.
15. Henry, J. (1984), Clinical diagnosis and management, 17th edition, pp 459-469.
16. Plum, F. & Siesjo, B. (1975), Anesthesiology, 42 ; 708-713.
17. Frey, A., Oechler, A., et al., (1989), Biological chemistry, 370 ; 997-1002.
18. De Bault, L., Henriquez, E., et al., (1981), In Vitro., 17 :; 488-494.
19. Reymond, A. (1981), Principles of neurology, 2nd edition, pp 11-15.
20. El-Mallakh, R. (1987), Am. Fam. Physician, 35 ; 112-118.
21. Lanningan, R., MacDonald, M., et al. (1980), J. Clin. Microbiol., 11 ; 324-328.
22. Teitz, N. (1982), Fundamentals of clinical chemistry, p. 369.
23. Killingsworth, L., Cooney, S., et al. (1980), Diagn. Med., March-April ; 23-29.
24. Calabrese, V. (1976), Vir. Med. Month. 103 ; 207-301.
25. Jameson, B. & Wells, D. (1972), N. Engl. J. Med., 286 ;1267-1271.
26. Watt, G. Zaraspe, G., et al. (1988), J. Infect. Dis., 158 ; 681-686.
27. Kaplan, A. (1989), Clinical chemistry, theory, analysis and correlation, 2nd edition, pp 1037-1038.
28. Schuller, E., Benabdallah, S., et al. (1987), Arch. Neurol., 44 ; 600-604.
29. Scopes, R. (1987), Protein purification, principles and practice, 2nd edition, p.280.
30. Finely, P. & Williams, R. (1983), Clin. Chem., 29 ; 126-129.
31. McIntosh, J. (1977), Clin. Chem., 23 ; 1939-1940.
32. Stahl, M. (1984), Clin. Chem., 30 ; 1878-1880.

33. Krystal, G., Lam,V. & Schreiber, W. (1989), Clin. Chem., 35 ; 860-864.
    Chatterley, S., Sun, T. & Lien, Y. (1991), J. Clin. Lab. Anal., 5 ; 168-174.
    Kaplan, A. (1989), Clinical chemistry, theory, analysis and correlation, 2nd edition, pp 784-793.

36. Lehninger, A., Nelson, D. & Cox, M. (1993), Principles of biochemistry, 2nd edition, pp 416-437.

37. Al-Mudhaffar, S. (1990), Biochemistry, part two (1), p. 421.

38. Stryer, L. (1995), Biochemistry, 4th edition, pp 483-498.

39. Stryer, L., ibid., P. 577.

40. Hanson, R. (1989), Biochem. Educ., 17, 86-92.

41. Srivastava, D. & Bernhard, S. (1987), Annu. Rev. Biophys. Biophys. Chem., 16 ; 175-204.

42. Pilkis, S. & Clause, T. (1991), Annu. Rev. Nutr., 11 ; 465-515.

43. Pilkis, S. & Granner, D. (1992), Ann. Rev. Physiol., 54 ; 885-909.

44. Dixon, M. & Webb, E. (1979), Enzymes, 3rd edition, p. 474.

45. Lehninger, A., Nelson, D. & Cox, M. (1993), Principles of biochemistry, 2nd edition, pp 391-392.

46. Bock, R. (1960), The Enzymes, 2nd edition, vol. 2, pp 33-38.

47. Kaplan, A. (1989), Clinical chemistry, theory, analysis and correlation, 2nd edition, pp 924-928.

48. Schultz, P. (1980), Science, 240 ; 426-433.

49. Scharpia, F. (1973), Adv. Cancer Res., 18 ; 77-153.

50. Iott, J. & Stang, J. (1980), Clin. Chem., 26 ; 1241-1250.

51. Wilhem, A. (1980), Artery, 8 ; 362-367.

52. Stefanini, M. (1985), Cancer, 55 ; 1931-1936.

53. Carola-Abella, P., Purez-Cuadrado, S., et al. (1982), Cancer, 49 ; 80-83.

54. Dimitrios, L. & Elefteria, E. (1989), Clin. Chem., 35 ; 844-848.

55. Giannoulakl, E., Tentas, C., et al. (1989), 35 ; 396-399.

56. Lubin, J., Cabello, B., et al. (1980), Am. J. Clin. Pathol., 73 ; 253-258.

57. Baghavan, N., Darm, J., et al. (1982), Arch. Pathol. Lab. Med., 106 ; 521-523.

58. Podlasek, S., McPherson, R. & Threatte, G. (1984), Clin. Chem., 30 ;266-270.

59. Otsu, N., Hirata, M., et al. (1985), Clin. Chem., 31 ; 318-320.

60. Zail, S. & Van den Hoek, A. (1977), Clin. Chim. Acta., 79 ; 15-19.

61. Siciliano, M., Bordelon-Rise, M.,et al. (1980), Cancer research, 40 ; 283-287.

62. Eldow, J., Huddleston, J., et al. (1971), Am. J .Obstet. Gynecol., 3 ; 360-364.

63. Van Bogaert,E.,De Peretti,E.&Villee,C.(1967), Am.J.Obstet.Gynecol.,98;919-923.

64. Anderson, G. & Kovacik, J. (1981), Proc. Natl. Acad. Sci. USA., 78 ; 3209-3213.

65.  Okuyama, M., Kuromiya, K., *et al.* (1982), Phys. Chem. Biol., 26 ; 43-47.
66.  Wickus, G. & Smith, M. (1984), Clin. Chem., 30 ; 11-17.
67.  Fujita, K. & Takeya, C. (1984), Clin. Chim. Acta., 140 ; 183-195.
68.  Matsumoto, K., Mori, Y., *et al.* (1986), Clin. Chem., 32 ; 1420-1422.
69.  Javed, M., Yousuf, F., *et al.* (1995),Comp. Biochem. Physiol. B. Biochem. Mol. Biol., 111 ; 27-34.
70.  Rotenberg, Z., Weinberger, E., *et al.* (1989), Clin. Chem., 35 ; 871-873.
71.  Marchat, L., Loiseau, P.,*et al.* (1995), Comp. Biochem. Physiol. B. Biochem. Mol. Biol., 109 ; 451-457.
72.  Usategue-Gomez, M., Wicks, R. & Warshaw, M. (1979), Clin. Chem., 25 ; 729-734.
73.  Kaplan, A. (1989), Clinical chemistry, theory, analysis and correlation, 2nd edition, p. 729.
74.  Howell, B., McClure, S. & Schaffer, R. (1979), Clin. Chem., 24 ; 828-831.
75.  Freer, D., Statland, B., *et al.* (1979), Clin. Chem., 25 ; 565-569.
76.  Buhl, S. & Jackson, K. (1978), Clin. Chem., 24 ; 828-831.
77.  Speigel, H., Symington, J.,*et al.* (1972), Clin. Chem., 7 ; 43-46.
78.  Teitz, N. (1982), Fundamentals of clinical chemistry, pp 655-660.
79.  Wacker, W., Ulmer, D. & Vallee, B. (1956), New Eng, J. Med., 255 ; 449-456.
80.  Burnett, R. (1980), Clin. Chem., 25 ; 644-646.
81.  Murthy, V. (1995), J. Clin. Lab. Anal., 9 ; 225-229.
82.  Hamilton, S., & Pardue, H. (1984), Clin. Chem., 30 ; 226-229.
83.  Buhl, S. & Jackson, K. (1977), Clin. Chem., 23 ; 1289-1294.
84.  Akrawi, B. (1985), MSc. thesis, college of science, university of Baghdad, pp 124-126.
85.  Teitz, N. (1982), Fundamentals of clinical chemistry, pp 577-578.
86.  Ananthanavayanah, P. & Ramakrishnan, S. (1979), Indian J. Med. Res., 68; 459-465.
87.  Modoveanue, N. & Tanasescu, O. (1972), Biochem. Exp. Biol., 10 ; 221-227.
88.  Morrison, J., Whybrew, D., *et al.* (1971). Am. J. Obstet. Gynecol., 110 ; 619-622.
89.  Henry, J. (1984), Clinical diagnosis and management, 17th edition, p. 1434.
90.  Feldman, W. (1975), Am. J. Dis. Child. 129 ; 77-80.
91.  Jain, M., Shah, A., *et al.* (1991), Indian. Pediatrics, 28 ; 369-374.
92.  Eastham, R. (1985), Biochemical values in clinical medicine, 17th edition, p. 206.
93.  Flasher, M., Wasserstrom, W., *et al.* (1981), Cancer, 47 ; 2654-2659.

94.  Lehninger, A., Nelson, D. & Cox, M. (1993), Principles of biochemistry, 2nd edition, p. 744.

95.   Kawamoto, M. (1994), Cancer, 73 ; 1836-1841.

96.   Mate, J., Duran, R., *et al.* (1993), Hepatogastroenterology, 40 ; 471-474.

97.   Plummer, G. (1971), An introduction to practical biochemistry, pp 157-158.

98.   Mancini, G., Larbonar, A. & Hareman, S. (1965), Immunochem., 2; 235-238.

99.   Arata, L. & Leonardi, A. (1988), Clin. Immunol. Immunopathol., 47 ; 10-18.

100.  Reymond, A. (1981), Principles of neurology, 2nd edition, p. 486.

101.  Scopes, R. (1987), Protein purification, principles and practice, 2nd edition, pp 11-12.

102.  Akrawi, B. (1985), MSc. thesis, college of science, university of Baghdad, pp 66-69.

103.  Mercer, D. (1978), Clin. Chem., 24 ; 480-482.

104.  Seigle, I. (1976), Biochemical calculations, 2nd edition, p. 273.

105.  Dixon, M. & Webb, E. (1979), Enzymes, 3rd edition, p. 273.

106.  Lehninger, A., Nelson, D. & Cox, M. (1993), Principles of biochemistry, 2nd edition, p. 212.

107.  Barrow, G. (1973), Physical Chemistry, 3rd edition, pp 424-430.

108.  Bernard, A., Fulpius, B., *et al.* (1977), J. Biol. Chem., 252 ; 4811-4830.

109.  Wielend, G. & Molinof, P. (1981), Life Sci., 29 ; 313-316.

110.  Waelbroeck, M., Van Obberghen, E. & De Meyts, P. (1979), J. Biol. Chem., 255 ; 7736-7740.

111.  Blumenthal, D. & Stull, J. (1982), Biochemistry, 21 ; 2386-2391.

112.  Laprte, D., Wierman, B. & Storm, D. (1980), Biochemistry, 19 ; 3814-3819.

113.  Lehninger, A., Nelson, D. & Cox, M. (1993), Principles of biochemistry, 2nd edition, p. 391.

www.ingramcontent.com/pod-product-compliance
Lightning Source LLC
Chambersburg PA
CBHW080829180526
45168CB00006B/2616